The Global Environment, Natural Resources, and Economic Growth

The Global Environment, Natural Resources, and Economic Growth

Alfred Greiner and Willi Semmler

OXFORD
UNIVERSITY PRESS
2008

OXFORD
UNIVERSITY PRESS

Oxford University Press, Inc., publishes works that further
Oxford University's objective of excellence
in research, scholarship, and education.

Oxford New York
Auckland Cape Town Dar es Salaam Hong Kong Karachi
Kuala Lumpur Madrid Melbourne Mexico City Nairobi
New Delhi Shanghai Taipei Toronto

With offices in
Argentina Austria Brazil Chile Czech Republic France Greece
Guatemala Hungary Italy Japan Poland Portugal Singapore
South Korea Switzerland Thailand Turkey Ukraine Vietnam

Copyright © 2008 by Oxford University Press, Inc.

Published by Oxford University Press, Inc.
198 Madison Avenue, New York, New York 10016

www.oup.com

Oxford is a registered trademark of Oxford University Press.

Library of Congress Cataloging-in-Publication Data
Greiner, Alfred.
The global environment, natural resources, and economic growth/
Alfred Greiner, Willi Semmler.
p. cm.
Includes bibliographical references and index.
ISBN 978-0-19-532823-3
1. Economic development—Environmental aspects. 2. Pollution—Economic aspects.
3. Natural resources—Management. I. Semmler, Willi. II. Title.
HD75.6.G745 2008
333.7—dc22q   2007047160

9 8 7 6 5 4 3 2 1

Printed in the United States of America
on acid-free paper

"We have not inherited the earth from our ancestors, we have only borrowed it from our children."

—Ancient Proverb

"Act so that the effects of your action are compatible with the permanence of genuine human life."

—Hans Jonas (1903–1993),
German-born philosopher,
taught at the New School, 1955–1976

Preface

Recently public attention has turned toward the intricate interrelation between economic growth and global warming. This book focuses on this nexus but broadens the framework to study this issue. Growth is seen as global growth, which affects the global environment and climate change. Global growth, in particular high economic growth rates, implies a fast depletion of renewable and nonrenewable resources. Thus the book deals with the impact of economic growth on the environment and the effect of the exhaustive use of natural resources as well as the reverse linkage. We thus address three interconnected issues: economic growth, environment and climate change, and renewable and nonrenewable resources. These three topics and the interrelationship among them need to be treated in a unified framework. In addition, not only intertemporal resource allocation but also the eminent issues relating to intertemporal inequities, as well as policy measures to overcome them, are discussed in the book. Yet more than other literature on global warming and resources, we study those issues in the context of modern growth theory. Besides addressing important issues in those areas we also put forward a dynamic framework that allows focus on the application of solution methods for models with intertemporal behavior of economic agents.

The material in this book has been presented by the authors at several universities and conferences. Chapters have been presented as lectures at Bielefeld University; Max Planck Institute for Demographic Research, Rostock; Sant'Anna School of Advanced Studies of Pisa, Itlay; University of Technology, Vienna; University of Aix-en-Provence; Bernard Schwartz Center for Economic Policy Analysis of the New School, New York; and Chuo University, Tokyo, Japan. Some chapters have also been presented at the annual conference of the Society of Computational Economics and the Society of Nonlinear Dynamics and Econometrics. We are grateful for comments by the participants of those workshops and conferences.

Some parts of the book are based on joint work with co-authors. Chapter 14 is based on the joint work of Almuth Scholl and Willi Semmler, and chapter 15 originated in the joint work of Malte Sieveking and Willi Semmler. We particularly want to thank Almuth Scholl and Malte Sieveking for allowing us to use this material here.

We are also grateful for discussions with and comments from Philippe Aghion, Toichiro Asada, Buz Brock, Graciela Chichilnisky, Lars Grüne, Richard Day, Ekkehard Ernst, Geoffrey Heal, James Ramsey, Hirofumi Uzawa, and colleagues of our universities. We thank Uwe Köller for research assistance and Gaby Windhorst for editing and typing the manuscript. Financial support from the Ministry of Education, Science and Technology of the State of Northrhine-Westfalia, Germany, and from the Bernard Schwartz Center for Economic Policy Analysis of the New School is gratefully acknowledged. Finally we want to thank numerous anonymous readers and Terry Vaughn and Catherine Rae at Oxford University Press, who have helped the book to become a better product.

Contents

Part II   Global Warming and Economic Growth

Part III   Depletion of Resources and Economic Growth   123

The Global Environment, Natural Resources, and Economic Growth

Introduction

The globalization of economic activities since the 1980s and 1990s, accelerated through free trade agreements, liberalized capital markets, and labor mobility, has brought into focus the issues related to global growth, resources, and environment. The industrialization in many countries in the past 100 years and the resource-based industrial activities have used up resources, mostly produced by poor and developing countries. The tremendous industrial growth in the world economy, particularly since World War II, and the current strong economic growth in some regions of the world, for example in Asia and some Latin American countries, have generated a high demand for specific inputs. Renewable as well as nonrenewable resources have been in high demand, and they are threatened with being depleted. In particular, the growing international demand for metals and energy derived from fossil fuels, as well as other natural resources, which are often extracted from developing countries, has significantly reduced the years to exhaustion for those resources.

It is true that technical progress has reduced the dependence of modern economies on natural resources, which is beneficial for their conservation, but this positive effect mostly holds for advanced economies producing with up-to-date technologies. Developing nations producing with older technologies usually do not have this advantage. In addition, several of those countries have experienced high growth rates over the past years. In particular, China and India have grown very fast over the past decades. These two countries alone comprise a population of more than two billion citizens, and the high growth rates in these countries have led to a dramatic increase in the demand for natural resources.

Whereas modern economies, like those in Western Europe and Japan, could reduce their dependence on nonrenewable resources, this does not necessarily hold for renewable resources. In particular, many oceans have been overfished for a long time. Current estimates assume that about 75 percent of the worldwide fish population is overfished. Although this problem has been realized by scientists and politicians, the short-run gains seem to be more important than conservation, leading to a severe threat to some fish species.

There is also an issue of inequity involved. An overwhelming fraction of resources, located in the South, are used up in the North, in

3

the industrialized countries, and the North has become the strongest polluter of the global environment. Many recent studies have confirmed that the emission of greenhouse gases is the main cause for global warming. Moreover, concerning intergenerational equity, current generations extensively use up resources and pollute the environment. Both produce negative externalities for future generations.

Indeed, not only does the environmental pollution strongly affect the current generation, but the environmental degradation affects future generations as well. It is true that as for the dependence on natural resources, technical progress has led to a more efficient use of technologies so that emissions of some pollutants have been reduced considerably. Indeed, in a great many regions in Europe and in the United States, for example, air pollution has been successfully reduced, leading to a cleaner environment. However, this does not hold for all types of emissions. In particular, emissions of greenhouse gases are at a high level and still increasing. Concerning greenhouse gas emissions, the high standard of living of modern Western societies makes these countries emit most of these gases, if measured per capita. Since the conference and protocol of Kyoto in 1997, the global change of the climate has become an important issue for academics as well as politicians. Although some countries had cast doubt on the fact that it is humankind that produces a global climate change, this question seems to have been answered now. There is vast evidence that the climate of the Earth is changing due to increases in greenhouse gases caused by human activities (see, for example, the report by Stern 2006, 2007, and the IPCC report 2007).

Although some may argue that to address and study those issues on global growth, environment, and resources, large-scale macro models may be needed. Yet when those models are solved through simulations, the mechanisms get blurred, and policy implication are not transparently derived. This book takes a different route. In the context of modern small-scale growth models, where the behavior of the agents and the framework are well defined, clear and coherent results are derived that may become useful guidelines for policy makers and practitioners.

The outline of the book reflects the discussed major issues. Part I deals with the environment and growth. We present models that incorporate the role of environmental pollution into modern growth models and derive optimal abatement activities as public policy. Part II models global climate change in the context of economic growth models. Policy implications are direct and transparent. Part III evaluates the use and overuse of nonrenewable and renewable resources in the context of intertemporal economic models. Aspects of global and intertemporal inequities as well as policy measures to overcome them are discussed in each part of the book.

PART I
The Environment and Economic Growth

1

Introduction and Overview

There are numerous economic models that study the interrelation between economic growth and the environment. We focus on a class of models in which economic activities lead to environmental degradation, and thus economic activity negatively affect the utility of households or the production activities of firms. This line of research goes back to Forster (1973) and was extended by Gruver (1976). Forster (1973), for example, studies a dynamic model of capital accumulation, the Ramsey growth model, with pollution as a byproduct of capital accumulation that can be reduced by abatement spending. In the long run, this model is characterized by a stationary state where all variables are constant unless exogenous shocks occur.

Another early contribution in environmental economics is the book by Mäler (1974), which can be considered as a classical contribution in this field. Mäler analyzes several aspects associated with environmental degradation in different frameworks, such as a general equilibrium model of environmental quality and an economic growth model incorporating the environment. But Mäler assumes a finite time horizon and is less interested in the long-run evolution of economies, in contrast to Forster (1973).

If one studies a growth model and intends to analyze the long-run evolution of economies, models with constant variables in the long run are rather unrealistic. With the publication of the papers by Romer (1986, 1990) the "new" or endogenous growth theory has become prominent. The major feature of models within this line of research is that the growth rate becomes an endogenous variable, the per capita income rises over time, and the government may affect growth through fiscal policy, for example. Concerning the forces that can generate ongoing growth, one can think of positive externalities associated with investment, the formation of human capital, or the creation of a stock of knowledge through R&D spending (for a survey, see Greiner et al. 2005).

Another type of model in endogenous growth theory assumes that the government can invest in productive public capital, which stimulates aggregate productivity. This approach goes back to Arrow and Kurz (1970), who presented exogenous growth models with that assumption in their book. The first model in which productive public spending leads to sustained per capita growth in the long run was presented by Barro (1990). In his model, productive public spending positively affects the marginal product of private capital and makes the long-run growth rate an endogenous variable. However, the assumption that public spending as a flow variable affects aggregate production activities is less plausible from an empirical point of view, as pointed out in a study by Aschauer (1989).

Futagami et al. (1993) have extended the Barro model by assuming that public capital as a stock variable shows positive productivity effects and then investigated whether the results derived by Barro are still valid given their modification of the model. However, the assumption made by these researchers implies that the model has transition dynamics, which does not hold for the model when public spending as a flow variable shows productive effects. In the latter case, the economy immediately jumps on the balanced growth path. The model presented by Futagami and colleagues is characterized by a unique balanced growth path, which is a saddle point. Although the questions of whether the long-run balanced growth path is unique and whether it is stable are important issues, they are not frequently studied in this type of research. Most of the contributions study growth and welfare effects of fiscal policy for a model on the balanced growth path.

As to the question of whether public spending can affect aggregate production possibilities at all, the empirical studies do not obtain unambiguous results. However, this is not too surprising because these studies often consider different countries over different time periods and the effect of public investment in infrastructure, for example, is likely to differ over countries and over time. A survey of the empirical studies dealing with that subject can be found in Pfähler et al. (1996), Sturm et al. (1998), Romp and de Haan (2005), and Semmler et al. (2007).

Problems of environmental degradation have also been studied in endogenous growth models. There exist many models dealing with environmental quality or pollution and endogenous growth (for a survey, see, for example, Smulders 1995 or Hettich 2000). Most of these models assume that pollution or the use of resources influences production activities either through affecting the accumulation of human capital or by directly entering the production function. Examples of that type of research are the publications by Bovenberg and Smulders (1995), Gradus and Smulders (1993), Bovenberg and de Mooij (1997), and Hettich (1998). The goal of these studies, then, is to analyze how different tax policies affect growth, pollution, and welfare in an economy. But as with the approaches already mentioned, most of these models do not have transition dynamics or the analysis is limited to the balanced growth path. An explicit analysis of the dynamics is often beyond the scope of these contributions. An exception is provided by the paper by Koskela et al. (2000), who study an overlapping generations model with a renewable resource that serves as a store of value and as an input factor in the production of the consumption good. They find that indeterminacy and cycles may result in their model, depending on the value of the intertemporal elasticity of consumption.

In part I we analyze a growth model where pollution only affects utility of a representative household but does not affect production activities directly through entering the aggregate production function. However, there is an indirect effect of pollution on output because we suppose that resources are used for abatement activities. Concerning pollution, we assume that it is an inevitable byproduct of production and can be reduced to a certain degree by investing in abatement activities. As to the growth rate, we suppose that it is determined endogenously and that public investment in a productive public capital stock brings about sustained long-run per capita growth. Thus we adopt that type of endogenous growth models that was initiated by Barro (1990), Futagami et al. (1993), and others as mentioned.

Our approach is closely related to the contributions by Smulders and Gradus (1996) and Bovenberg and de Mooij (1997), who are interested in growth and welfare effects of fiscal policy affecting the environment but do not explicitly study the dynamics of their models. Concerning the structure, our model is similar to the one presented by Bovenberg and de Mooij (1997) with the exception that we assume that public capital as a stock enters the aggregate production function, whereas Bovenberg and de Mooij assume that public investment as a flow has positive effects on aggregate production.

In chapter 2 we present a simple variant of an economic model with environmental pollution and productive public capital. This model will be analyzed assuming a logarithmic utility function. Chapter 3 studies both growth and welfare effects of fiscal policy. In particular, we analyze how the long-run balanced growth rate reacts to fiscal policy and to the introduction of a less polluting technology. Further, we study the effects of fiscal policy, taking into account transition dynamics, and we analyze welfare effects of fiscal policy on the environment on the balanced growth path as well as the social optimum. In chapter 4 we generalize our model and allow for a more general isoelastic utility function. The goal, then, is to give an explicit characterization of the dynamic behavior resulting from more general assumptions. An extension of the model is presented in chapter 5 , where we assume that environmental pollution as a stock negatively affects utility of the household. In this variation of the model, we consider three different scenarios: first, we analyze a scenario with a constant stock of pollution; second, we study a scenario with an improving environmental quality; and finally, we analyze a scenario in which environmental pollution grows at the same positive rate as all other endogenous variables.

2

The Basic Economic Model

We consider a decentralized economy with a household sector, a productive sector, and the government (see Greiner 2005a). First, we describe the household sector. For reasons of simplicity we presume here the household's preferences to be logarithmic in consumption and pollution.

2.1 THE HOUSEHOLD SECTOR

The household sector in our economy consists of many identical households, which are represented by one household. The goal of this household is to maximize a discounted stream of utility arising from consumption $C(t)$ over an infinite time horizon subject to its budget constraint:

$$\max_{C(t)} \int_0^\infty e^{-\rho t} V(t) dt, \qquad (2.1)$$

with $V(t)$ the instantaneous subutility function that depends positively on the level of consumption and negatively on effective pollution, $P_E(t)$. $V(t)$ takes the logarithmic form

$$V(t) = \ln C(t) - \ln P_E(t), \qquad (2.2)$$

with ln giving the natural logarithm.[1] ρ in (2.1) is the subjective discount rate. Later in the book, at various places, we discuss further the importance of the discount rate for the solution of our models. Here it may suffice to refer the reader to an important recent work on the discount rate; see Weitzman (2007a,b).

The budget constraint for the household is given by[2]

$$\dot{K} = (w + rK)(1 - \tau) - C. \qquad (2.3)$$

The budget constraint (2.3) states that the individual has to decide how much to consume and how much to save, thus increasing consumption

---

[1] For a survey of how to incorporate pollution in the utility function, see Smulders (1995), pp. 328–29.
[2] In what follows we suppress the time argument if no ambiguity arises.

possibilities in the future.[3] The depreciation of physical capital is assumed to equal zero.

The wage rate is denoted by w. The labor supply L is constant, supplied inelastically, and we normalize $L \equiv 1$. r is the return to per capita capital K, and $\tau \in (0,1)$ gives the income tax rate.

To derive necessary conditions we formulate the current-value Hamiltonian function as

$$\mathcal{H}(\cdot) = \ln C - \ln P_E + \lambda(-C + (w + rK)(1 - \tau)), \qquad (2.4)$$

with λ the costate variable. The necessary optimality conditions are given by

$$\lambda = C^{-1}, \qquad (2.5)$$

$$\dot{\lambda}/\lambda = \rho - r(1 - \tau), \qquad (2.6)$$

$$\dot{K} = -C + (w + rK)(1 - \tau). \qquad (2.7)$$

Because the Hamiltonian is concave in C and K jointly, the necessary conditions are also sufficient if in addition the transversality condition at infinity $\lim_{t \to \infty} e^{-\rho t} \lambda(t) K(t) = 0$ is fulfilled. Moreover, strict concavity in C also guarantees that the solution is unique (see the appendix and, for more details, Seierstad and Sydsaeter [1987], pp. 234–35).

2.2 THE PRODUCTIVE SECTOR

The productive sector in our economy consists of many identical firms that can be represented by one firm. The latter behaves competitively and chooses inputs to maximize profits.

As to pollution P, we suppose that it is a byproduct of aggregate production Y. In particular, we assume that

$$P(t) = \varphi Y(t), \qquad (2.8)$$

with $\varphi = const. > 0$. Thus, we follow the line invited by Forster (1973) and worked out in more detail by Luptacik and Schubert (1982).

Effective pollution P_E, which affects utility of the household, is that part of pollution that remains after investing in abatement activities A. This means that abatement activities reduce pollution but cannot eliminate it completely. As to the modeling of effective pollution, we follow Gradus and Smulders (1993) and Lighthart and van der Ploeg (1994) and make the following specification:

$$P_E = \frac{P}{A^\beta}, \quad 0 < \beta \leq 1. \qquad (2.9)$$

---

[3] The dot over a variable gives the derivative with respect to time.

The limitation $\beta \leq 1$ ensures that a positive growth rate of aggregate production goes along with an increase in effective pollution, $\beta < 1$, or leaves effective pollution unchanged, $\beta = 1$. We make that assumption because we think it is realistic to assume that higher production also leads to an increase in pollution, although at a lower rate because of abatement. In looking at the world economy, that assumption is certainly justified. But it should also be noted that for $\beta = 1$ sustained output growth goes along with a constant level of effective pollution, which will be seen in detail in the next section. Further, we posit that $A^\beta > 1$ holds such that effective pollution is smaller than pollution without abatement, which is in a way an obvious assumption.

Pollution is taxed at the rate $\tau_p > 0$, and the firm takes into account that one unit of output causes φ units of pollution, for which it has to pay $\tau_p \varphi$ per unit of output. The per capita production function is given by

$$Y = K^\alpha H^{1-\alpha} L^{1-\alpha} \equiv K^\alpha H^{1-\alpha}, \qquad (2.10)$$

with H denoting the stock of productive public capital and $\alpha \in (0,1)$ giving the per capita capital share. Recall that K denotes per capita capital and L is normalized to one.

Assuming competitive markets and taking public capital as given, the first-order conditions for a profit maximum are obtained as

$$w = (1 - \tau_p \varphi)(1 - \alpha) K^\alpha H^{1-\alpha}, \qquad (2.11)$$

$$r = (1 - \tau_p \varphi) \alpha K^{\alpha-1} H^{1-\alpha}. \qquad (2.12)$$

2.3 THE GOVERNMENT

The government in our economy uses resources for abatement activities $A(t)$ that reduce total pollution. Abatement activities $A \geq 0$ are financed by the tax revenue coming from the tax on pollution, that is, $A(t) = \eta \tau_p P(t)$, with $\eta > 0$. If $\eta < 1$, not all of the pollution tax revenue is used for abatement activities and the remaining part is spent for public investment in the public capital stock I_p, $I_p \geq 0$, in addition to the tax revenue resulting from income taxation. For $\eta > 1$ a certain part of the tax revenue resulting from the taxation of income is used for abatement activities in addition to the tax revenue gained by taxing pollution. As to the interpretation of public capital, one can think of infrastructure capital, for example. However, one could also interpret public capital in a broader sense so that it also includes human capital, which is built up as a result of public education.

It should be mentioned that there are basically two approaches in the economics literature to model abatement. The first assumes that private firms engage in abatement (see Bovenberg and de Mooij 1997 or Bovenberg and Smulders 1995). In the second approach, which we follow here, abatement spending is financed by the government (see Lighthart and van der Ploeg 1994 or Nielsen et al. 1995).

The government in our economy runs a balanced budget at any moment in time. Thus, the budget constraint of the government is written as

$$I_p = \tau_p P(1 - \eta) + \tau(w + rK). \tag{2.13}$$

The evolution of public capital is described by

$$\dot{H} = I_p, \tag{2.14}$$

where for simplicity we again assume that there is no depreciation of public capital.

2.4 EQUILIBRIUM CONDITIONS AND THE BALANCED GROWTH PATH

Combining the budget constraint of the government and the equation describing the evolution of public capital over time, the accumulation of public capital can be written as

$$\dot{H} = -\eta\varphi\tau_p K^\alpha H^{1-\alpha} + \tau_p\varphi K^\alpha H^{1-\alpha} + \tau(w + rK)$$
$$= K^\alpha H^{1-\alpha}(\varphi\tau_p(1 - \eta) + (1 - \varphi\tau_p)\tau), \tag{2.15}$$

where we have used (2.11) and (2.12).

To obtain the other differential equations describing our economy, we note that the growth rate of private consumption is obtained from (2.5) and (2.6), with r taken from (2.12) and where we have used $\dot{P}_E/P_E = (1 - \beta)\dot{Y}/Y$. Using (2.11) and (2.12), $\dot{K}/K$ is obtained from (2.7). It should be noted that the accumulation of public capital, which is positive for $I_p > 0$, is the source of sustained economic growth in our model and makes the growth rate an endogenous variable.

Thus the dynamics of our model are completely described by the following differential equation system:

$$\frac{\dot{C}}{C} = -\rho + (1 - \tau)(1 - \varphi\tau_p)\alpha\left(\frac{H}{K}\right)^{1-\alpha}, \tag{2.16}$$

$$\frac{\dot{K}}{K} = -\frac{C}{K} + \left(\frac{H}{K}\right)^{1-\alpha}(1 - \varphi\tau_p)(1 - \tau), \tag{2.17}$$

$$\frac{\dot{H}}{H} = \left(\frac{H}{K}\right)^{-\alpha}(\varphi\tau_p(1 - \eta) + (1 - \varphi\tau_p)\tau). \tag{2.18}$$

The initial conditions $K(0)$ and $H(0)$ are given and fixed, and $C(0)$ can be chosen freely by the economy. Further, the transversality condition $\lim_{t\to\infty} e^{-\rho t} K(t)/C(t) = 0$ must be fulfilled.[4]

In the following, we first examine our model as to the existence and stability of a balanced growth path (BGP). To do so, we define a BGP.

Definition 1 *A balanced growth path (BGP) is a path such that $\dot{C}/C = \dot{K}/K = \dot{H}/H \equiv g > 0$ holds, with g constant and C, K, and H strictly positive. A balanced growth path is sustainable if $\dot{V} > 0$ holds.*

This definition shows that on a BGP the growth rates of economic variables are positive and constant over time. Notice that aggregate output and pollution grow at the same rate on the BGP. This implies that effective pollution is not constant in the long run (unless $\beta = 1$ holds). Nevertheless, one may say that the BGP is sustainable if one adopts the definition given in Byrne (1997), which is done in our definition. There, sustainable growth is given if instantaneous utility grows over time, that is, if $\dot{V}$ is positive. For our model with logarithmic utility, this is automatically fulfilled on the BGP because $\dot{V} = \dot{C}/C - \dot{P}_E/P_E = \dot{C}/C - (1-\beta)\dot{Y}/Y = \beta g > 0$ holds.

To analyze the model further, we first have to perform a change of variables. Defining $c = C/K$ and $h = H/K$ and differentiating these variables with respect to time, we get $\dot{c}/c = \dot{C}/C - \dot{K}/K$ and $\dot{h}/h = \dot{H}/H - \dot{K}/K$. A rest point of this new system then corresponds to a BGP of our original economy where all variables grow at the same constant rate. The system describing the dynamics around a BGP is given by

$$\dot{c} = c\big(c - \rho - (1-\alpha)(1-\tau)(1-\varphi\tau_p)h^{1-\alpha}\big), \qquad (2.19)$$

$$\dot{h} = h\big(c - h^{1-\alpha}(1-\varphi\tau_p)(1-\tau) + h^{-\alpha}(\varphi\tau_p(1-\eta) + (1-\varphi\tau_p)\tau)\big). \quad (2.20)$$

Concerning a rest point of system (2.19) and (2.20), note that we only consider interior solution. That means that we exclude the economically meaningless stationary point $c = h = 0$. As to the uniqueness and stability of a BGP, we can state the following proposition.

Proposition 1 *Assume that $\tau_p\varphi < 1$ and $(1 - \tau_p\varphi)\tau + (1 - \eta)\tau_p\varphi > 0$. Then there exists a unique BGP which is saddle point stable.*

Proof: To prove that proposition we first calculate $c^\star$ on a BGP,[5] which is obtained from $\dot{h}/h = 0$ as

$$c^\star = h^{1-\alpha}(1 - \varphi\tau_p)(1 - \tau) - h^{-\alpha}(\varphi\tau_p(1-\eta) + (1-\varphi\tau_p)\tau).$$

---

[4] Note that (2.5) yields $\lambda = 1/C$.
[5] We denote values on the BGP by $\star$.

Inserting c^* in (2.19) gives after some modifications

$$f(\cdot) \equiv \dot{c}/c = -\rho + (1 - \tau)(1 - \varphi\tau_p)\alpha h^{1-\alpha} - h^{-\alpha}(\varphi\tau_p(1 - \eta) + (1 - \varphi\tau_p)\tau),$$

with $\lim_{h\to 0} f(\cdot) = -\infty$ (for $I_p > 0$) and $\lim_{h\to\infty} f(\cdot) = \infty$. A rest point for $f(\cdot)$, that is, a value for h such that $f(\cdot) = 0$ holds, then gives a BGP for our economy. Further, we have

$$\partial f(\cdot)/\partial h = (1 - \tau)(1 - \varphi\tau_p)(1 - \alpha)\alpha h^{-\alpha} + \alpha h^{-\alpha-1}(\varphi\tau_p(1 - \eta)$$
$$+ (1 - \varphi\tau_p)\tau) > 0,$$

for $I_p > 0$. Note that on a BGP $\dot{H}/H > 0$ must hold, implying $I_p > 0$ and thus $\varphi\tau_p(1 - \eta) + (1 - \varphi\tau_p)\tau > 0$. $\partial f/\partial h > 0$ for h such that $f(\cdot) = 0$ means that $f(\cdot)$ cannot intersect the horizontal axis from above. Consequently, there exists a unique h^* such that $f(\cdot) = 0$ and, therefore, a unique BGP.

The saddle point property is shown as follows. Denoting with J the Jacobian matrix of (2.19) and (2.20) evaluated at the rest point we first note that det $J < 0$ is a necessary and sufficient condition for saddle point stability, that is, for one negative and one positive eigenvalue. The Jacobian in our model can be written as

$$J = \begin{bmatrix} c & -ch^{-\alpha}(1 - \alpha)^2(1 - \tau)(1 - \tau_p\varphi) \\ h & -\upsilon \end{bmatrix},$$

with

$$\upsilon = (1 - \alpha)h^{1-\alpha}(1 - \tau)(1 - \tau_p\varphi) + \alpha h^{-\alpha}(\varphi\tau_p(1 - \eta) + (1 - \varphi\tau_p)\tau).$$

The determinant can be calculated as

$$\det J = -ch\alpha\big(h^{-\alpha-1}(\varphi\tau_p(1 - \eta) + (1 - \varphi\tau_p)\tau)$$
$$+ (1 - \alpha)h^{-\alpha}(1 - \tau)(1 - \tau_p\varphi)\big) < 0.$$

Thus, the proposition is proved. □

Proposition 1 states that our model is both locally and globally determinate, that is, there exists a unique value for $c(0)$ such that the economy converges to the unique BGP in the long run. Note that we follow Benhabib and Perli (1994) and Benhabib et al. (1994) concerning the definition of local and global determinacy. According to that definition, local determinacy is given if there exist unique values for the variables that are not predetermined but can be chosen at $t = 0$, such that the economy converges to the BGP in the long run. If there exists a continuum of values for the variables that can be chosen at time $t = 0$, so that the economy asymptotically converges to the BGP, the model reveals local indeterminacy.

Global indeterminacy arises if there exists more than one BGP and the variables that are not predetermined at time $t = 0$ may be chosen to place the economy on the attracting set of either of the BGPs. In this case, the initial choice of the variables, which can be chosen at time $t = 0$, affects not only the transitional growth rates but also the long-run growth rate on the BGP. If the long-run BGP is unique, the economy is said to be globally determinate.

The assumption $(1 - \varphi\tau_p) > 0$ is necessary for a positive growth rate of consumption and is sufficient for a positive value of $c^\star$.[6] The second assumption $(\varphi\tau_p(1 - \eta) + (1 - \varphi\tau_p\tau) > 0$ must hold for a positive growth rate of public capital. Note that $(\varphi\tau_p(1 - \eta) + (1 - \varphi\tau_p\tau)) = I_p/Y$, stating that the second assumption in proposition 1 means that on a BGP the ratio of public investment to GDP must be positive.

---

[6] This is realized if $c^\star$ is calculated from $\dot{c}/c = 0$ as $c^\star = \rho + (1 - \alpha)(1 - \tau)(1 - \tau_p\varphi)h^{1-\alpha}$.

3

Growth and Welfare Effects of Fiscal Policy

In the last chapter we demonstrated that there exists a unique BGP under slight additional assumptions. Thus, our model including the transition dynamics is completely characterized. In this chapter we analyze how the growth rate and welfare in our economy react to fiscal policy. The first will be done for the model on the BGP and taking into account transition dynamics, and the latter is done for the model on the BGP.

3.1 GROWTH EFFECTS OF FISCAL POLICY ON THE BGP

Before we analyze growth effects of fiscal policy, we study effects of introducing a less polluting production technology, that is, the impact of a decline in φ.

The balanced growth rate, which we denote by g, is given by (2.18) as

$$g = \dot{H}/H = (h)^{-\alpha} \left(\varphi\tau_p(1 - \eta) + (1 - \varphi\tau_p)\tau \right).$$

Differentiating g with respect to φ gives

$$\frac{\partial g}{\partial \varphi} = h^{-\alpha}\tau_p(1 - \eta - \tau) - (\varphi\tau_p(1 - \eta) + (1 - \varphi\tau_p)\tau)\alpha h^{-\alpha-1}\frac{\partial h}{\partial \varphi}.$$

$\partial h/\partial \varphi$ is obtained by implicit differentiation from $f(\cdot) = 0$ (from the proof of proposition 1) as

$$\frac{\partial h}{\partial \varphi} = \frac{\tau_p(1 - \eta - \tau) + \tau_p(1 - \tau)\alpha h}{\alpha(1 - \tau)(1 - \varphi\tau_p)(1 - \alpha) + \alpha h^{-1}(\varphi\tau_p(1 - \eta) + (1 - \varphi\tau_p)\tau)}.$$

For $(1 - \eta - \tau) = 0$ we get $\partial g/\partial \varphi < 0$. To get results for $(1 - \eta - \tau) \neq 0$ we insert $\partial h/\partial \varphi$ in $\partial g/\partial \varphi$. That gives

$$\frac{\partial g}{\partial \varphi} = h^{-\alpha}\tau_p(1 - \eta - \tau)$$

$$\cdot \left(1 - \frac{(\varphi\tau_p(1 - \eta) + (1 - \varphi\tau_p)\tau)[(1 - \eta - \tau) + h\alpha(1 - \tau)]}{(1 - \eta - \tau)\left[\begin{array}{c}(\varphi\tau_p(1 - \eta) + (1 - \varphi\tau_p)\tau) \\ +h(1 - \tau)(1 - \tau_p\varphi)(1 - \alpha)\end{array}\right]} \right).$$

17

From that expression, it can be seen that the expression in brackets is always positive for $(1 - \eta - \tau) < 0$ such that $\partial g / \partial \varphi < 0$. For $(1 - \eta - \tau) > 0$ it is immediately seen that

$$\frac{\partial g}{\partial \varphi} > = < 0 \Leftrightarrow (1 - \varphi \tau_p)(1 - \alpha)(1 - \eta - \tau)$$

$$> = < \alpha (\varphi \tau_p (1 - \eta) + (1 - \varphi \tau_p)\tau),$$

which simplifies to

$$\frac{\partial g}{\partial \varphi} > = < 0 \Leftrightarrow (1 - \eta)(1 - \alpha) > = < \varphi \tau_p (1 - \eta) + (1 - \varphi \tau_p)\tau.$$

The right-hand side of that expression is equivalent to I_P / Y. Thus we have proved the following proposition.

Proposition 2 If $(1 - \eta - \tau) \leq 0$, the use of a less polluting technology raises the balanced growth rate. For $(1 - \eta - \tau) > 0$, the use of a less polluting technology raises (leaves unchanged, lowers) the balanced growth rate if

$$\frac{I_p}{Y} > (=, <)(1 - \alpha)(1 - \eta).$$

To interpret that result we first note that a cleaner production technology (i.e., a lower φ) shows two different effects: on one hand, it implies that fewer resources are needed for abatement activities, leaving more resources for public investment. That effect leads to a higher ratio H/K, thus raising the marginal product of private capital r in (2.12). That is, the return on investment rises. Further, a less polluting technology implies that the firm has to pay less pollution taxes (the term $(1 - \tau_p \varphi)$ rises), which also has a stimulating effect on r, which can be seen from (2.12) and which also raises the incentive to invest. On the other hand, less pollution implies that the tax revenue resulting from the taxation of pollution declines and thus so does productive public spending. That effect tends to lower the ratio H/K and, therefore, the marginal product of private capital. This tends to lower the balanced growth rate.

If $\eta \geq 1 - \tau$, that is, if much of the pollution tax is used for abatement activities, a cleaner technology always raises the balanced growth rate. In that case, the negative growth effect of a decline in the pollution tax revenue is not too strong because most of that revenue is used for abatement activities that are nonproductive anyway. If, however, $\eta < 1 - \tau$, that is, a good deal of the pollution tax is used for productive government spending, a cleaner technology may either raise or lower economic growth. It increases the balanced growth rate if the share of public investment per GDP is larger than a constant that positively depends on the elasticity of aggregate output with respect to public capital and negatively on η, and vice versa.

Let us next study growth effects of varying the income tax rate. The next proposition demonstrates that a rise in that tax may have positive or negative growth effects and that there exists a growth-maximizing income tax rate.

Proposition 3 *Assume that there exists an interior growth-maximizing income tax rate. Then this tax rate is given by*

$$\tau = (1 - \alpha) - \alpha \varphi \tau_p (1 - \eta)/(1 - \varphi \tau_p).$$

Proof: To calculate growth effects of varying τ we take the balanced growth rate g from (2.18) and differentiate it with respect to that parameter. Doing so gives

$$\frac{\partial g}{\partial \tau} = h^{-\alpha}(1 - \tau_p \varphi) \left(1 - \frac{\alpha((1 - \tau_p \varphi)\tau + (1 - \eta)\tau_p \varphi)}{1 - \tau_p \varphi} \frac{\partial h}{\partial \tau} \frac{1}{h} \right),$$

where $\partial h/\partial \tau$ is obtained by implicit differentiation from $f(\cdot) = 0$ leading to

$$\frac{\partial h}{\partial \tau} = \frac{(1 - \varphi \tau_p)(1 + \alpha h)h}{h(1 - \tau)(1 - \varphi \tau_p)(1 - \alpha)\alpha + \alpha((1 - \tau_p \varphi)\tau + (1 - \eta)\tau_p \varphi)}.$$

Inserting $\partial h/\partial \tau$ in $\partial g/\partial \tau$ we get

$$\frac{\partial g}{\partial \tau} = h^{-\alpha}(1 - \tau_p \varphi)$$

$$\times \left(1 - \frac{((1 - \tau_p \varphi)\tau + (1 - \eta)\tau_p \varphi)(1 + \alpha h)}{h(1 - \tau)(1 - \varphi \tau_p)(1 - \alpha) + ((1 - \tau_p \varphi)\tau + (1 - \eta)\tau_p \varphi)} \right),$$

showing that

$$\frac{\partial g}{\partial \tau} > = < 0 \Leftrightarrow (1 - \tau)(1 - \varphi \tau_p)(1 - \alpha) > = < \alpha((1 - \tau_p \varphi)\tau + (1 - \eta)\tau_p \varphi).$$

Solving for τ gives

$$\frac{\partial g}{\partial \tau} > = < 0 \Leftrightarrow \tau < = > (1 - \alpha) - \alpha \varphi \tau_p (1 - \eta)/(1 - \varphi \tau_p)$$

That shows that the balanced growth rate rises with increases in τ as long as τ is smaller than the expression on the right-hand side, which is constant. $\square$

Proposition 3 shows that the growth-maximizing income tax rate does not necessarily equal zero in our model, which was to be expected because the government finances productive public spending with the tax revenue. There are two effects going along with variations of the income tax rate: on one hand, a higher income tax lowers the marginal product of private capital and, therefore, is a disincentive for investment.

On the other hand, the government finances productive public spending with tax revenue, leading to a rise in the ratio H/K, which raises the marginal product of private capital r and has, as a consequence, a positive effect on economic growth. However, boundary solutions, that is, $\tau_K = 0$ or $\tau_K = 1$, cannot be excluded. Whether there exists an interior or a boundary solution for the growth-maximizing capital income tax rate depends on the numerical specification of the parameters φ, τ_p, and η. Only for $\varphi\tau_p = 0$ or $\eta = 1$ is the growth-maximizing tax rate always in the interior of $(0, 1)$ and equal to the elasticity of aggregate output with respect to public capital.

Concerning the relation between the tax on pollution and the growth-maximizing income tax rate, we see that it negatively varies with the latter if $\eta < 1$. For $\eta > 1$ the growth maximizing income tax rate is the higher the higher the tax on pollution τ_p. The interpretation of that result is as follows: if $\eta < 1$, the government uses a part of the pollution tax revenue for the creation of public capital, which has positive growth effects. Increasing the tax on pollution implies that a part of the additional tax revenue is used for productive investment in public capital. Consequently, the income tax rate can be reduced without having negative growth effects. It should be noticed that a decrease in the income tax rate shows an indirect positive growth effect because it implies a reallocation of private resources from consumption to investment. In contrast to that, if $\eta > 1$ the whole pollution tax revenue is used for abatement activities. Raising the pollution tax rate in that situation implies that the additional tax revenue is used only for abatement activities but not for productive public spending. Consequently, the negative indirect growth effect of a higher pollution tax (through decreasing the return on capital r) must be compensated by an increase in the income tax rate. Note that the latter also has a negative indirect growth effect but that one is dominated in this case by the positive direct growth effect of higher productive public spending.

We next analyze long-run growth effects of a rise in the pollution tax rate. The result is summarized in the following proposition.

Proposition 4 *For $(1 - \eta - \tau) \leq 0$, a rise in the pollution tax rate always lowers the balanced growth rate. If $(1 - \eta - \tau) > 0$, the pollution tax rate maximizing the balanced growth rate is determined by*

$$\tau_p = \left(\frac{1}{\varphi}\right)\left(\frac{1 - \eta - \tau - \alpha(1 - \eta)}{1 - \eta - \tau}\right),$$

which is equivalent to

$$\frac{I_p}{Y} = (1 - \alpha)(1 - \eta).$$

Proof: To calculate growth effects of varying τ_p we take the balanced growth rate g again from (2.18) and differentiate it with respect to that parameter. Doing so gives

$$\frac{\partial g}{\partial \tau_p} = h^{-\alpha} \varphi (1 - \eta - \tau)$$

$$\cdot \left(1 - \frac{(\varphi \tau_p (1 - \eta) + (1 - \varphi \tau_p) \tau)[(1 - \eta - \tau) + h\alpha(1 - \tau)]}{(1 - \eta - \tau) \left[\begin{array}{c} (\varphi \tau_p (1 - \eta) + (1 - \varphi \tau_p) \tau) \\ + h(1 - \tau)(1 - \tau_p \varphi)(1 - \alpha) \end{array} \right]} \right).$$

From that expression, it can be seen that the expression in brackets is always positive for $(1 - \eta - \tau) < 0$ such that $\partial g / \partial \tau_p < 0$. For $(1 - \eta - \tau) = 0$ the result can directly be seen by multiplying out the expression above. For $(1 - \eta - \tau) > 0$, it is seen that

$$\frac{\partial g}{\partial \tau_p} > = < 0 \Leftrightarrow (1 - \varphi \tau_p)(1 - \alpha)(1 - \eta - \tau)$$

$$> = < \alpha (\varphi \tau_p (1 - \eta) + (1 - \varphi \tau_p) \tau),$$

which simplifies to

$$\frac{\partial g}{\partial \tau_p} > = < 0 \Leftrightarrow \tau_p <=> \left(\frac{1}{\varphi} \right) \left(\frac{1 - \eta - \tau - \alpha(1 - \eta)}{1 - \eta - \tau} \right)$$

and is equivalent to

$$\frac{\partial g}{\partial \tau_p} > = < 0 \Leftrightarrow \frac{I_p}{Y} <=> (1 - \eta)(1 - \alpha).$$

Thus, the proposition is proved.   □

The interpretation of that result is straightforward. An increase in the pollution tax rate always lowers the balanced growth rate if $(1 - \eta - \tau) \leq 0$. In that case, too much of the additional tax revenue (gained through the increase in τ_p) goes in abatement activities so that the positive growth effect of a higher pollution tax revenue (i.e., the increase in the creation of the stock of public capital) is dominated by the negative indirect effect of a reduction of the rate of return to physical capital r. The latter effect namely implies a reallocation of private resources from investment to consumption, which reduces economic growth. For $(1 - \eta - \tau) > 0$, however, there exists a growth-maximizing pollution tax rate.[1] In that case, the pollution tax has to be set such that public investment per GDP equals the elasticity of aggregate output with respect to public capital multiplied with that share of the pollution tax revenue not used for abatement activities but for productive public spending.

Further, notice that the growth-maximizing value of τ_p[2] is the higher the lower the amount of pollution tax revenue used for abatement activities. In the limit $(\eta = 0)$ we get the same result as in the study by Futagami et al. (1993) where the growth-maximizing share of public investment per GDP equals the elasticity of aggregate output with respect to public capital.

Also note that the conditions for a positive growth effect of an increase in the pollution tax rate are just reverse to the conditions that must be fulfilled such that the introduction of a less polluting technology raises economic growth.

3.2 GROWTH EFFECTS ON THE TRANSITION PATH

In this section we study how the growth rates of consumption and public and private capital react to a change in the income tax and pollution tax rate, taking into account transition dynamics. To do this we proceed as follows. We assume that initially the economy is on the BGP when the government changes the tax rates at time $t = 0$, and then we characterize the transition path to the new BGP, which is attained in the long run.

First, we consider the effects of an increase in the income tax rate τ. To do this we state that the $\dot{c} = 0$ and $\dot{h} = 0$ isoclines are given by

$$c \mid_{\dot{c}=0} = \rho + (1 - \alpha)(1 - \tau)(1 - \tau_p\varphi)h^{1-\alpha}, \qquad (3.1)$$

$$c \mid_{\dot{h}=0} = h^{1-\alpha}(1 - \tau)(1 - \tau_p\varphi) - h^{-\alpha}((1 - \tau_p\varphi)\tau + (1 - \eta)\tau_p\varphi). \qquad (3.2)$$

Calculating the derivative dc/dh, it can easily be seen that the $\dot{h} = 0$ isocline is steeper than the $\dot{c} = 0$ isocline. Further, for the $\dot{c} = 0$ isocline we have $c = \rho$ for $h = 0$ and $c \to \infty$ for $h \to \infty$. For the $\dot{h} = 0$ isocline we have $c \to -\infty$ for $h \to 0$, $c = 0$ for $h = ((1 - \tau_p\varphi)\tau + (1 - \eta)\tau_p\varphi)/((1 - \tau)(1 - \tau_p\varphi))$ and $c \to \infty$ for $h \to \infty$. This shows that there exists a unique $(c^\star, h^\star)$ where the two isoclines intersect.

If the income tax rate is increased, it can immediately be seen that the $\dot{h} = 0$ isocline shifts to the right and the $\dot{c} = 0$ isocline turns right with

---

[1] But it must kept in mind that $1 - \tau_p\varphi > 0$ must hold so that a BGP exists. Therefore, the boundary condition $\tau_p = \varphi^{-1} - \bar{\epsilon}, \bar{\epsilon} > 0$, cannot be excluded.

[2] Note that I_p/Y positively varies with τ_p for $(1 - \eta - \tau) > 0$.

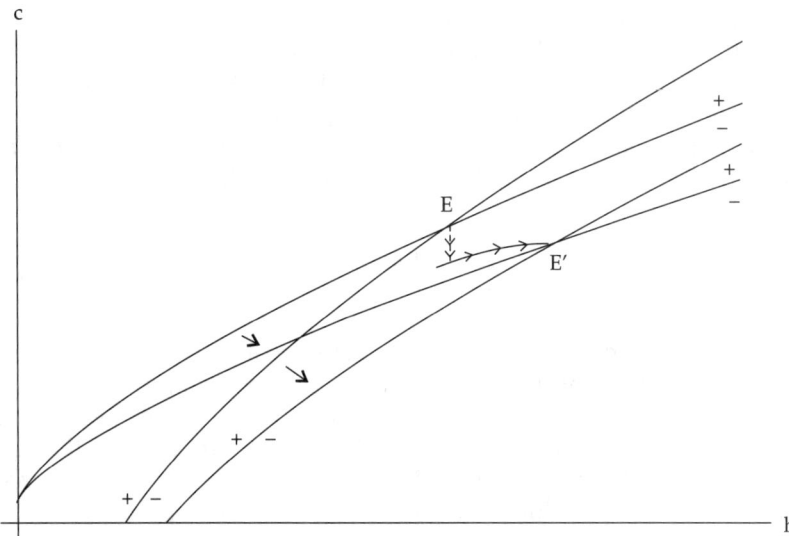

Figure 3.1 Effect of an increase in the income tax rate.

$c = \rho$ for $h = 0$ remaining unchanged. This means that on both the new $\dot{c} = 0$ and the new $\dot{h} = 0$ isocline any given h goes along with a lower value of c compared to the isoclines before the tax rate increase. This implies that the increase in the income tax rate raises the long-run value $h^\star$ and may reduce or raise the long-run value of $c^\star$. Further, the capital stocks K and H are predetermined variables that are not affected by the tax rate increase at time $t = 0$. These variables react only gradually. This implies that $\partial h(t = 0, \tau)/\partial \tau = 0$. To reach the new steady state[3] $(c^\star, h^\star)$ the level of consumption adjusts and jumps to the stable manifold implying $\partial c(t = 0, \tau)/\partial \tau < 0$, as shown in figure 3.1.

Over time both c and h rise until the new BGP is reached at $(c^\star, h^\star)$. That is, we get $\dot{c}/c = \dot{C}/C - \dot{K}/K > 0$ and $\dot{h}/h = \dot{H}/H - \dot{K}/K > 0$, implying that on the transition path the growth rates of consumption and public capital are larger than that of private capital for all $t \in [0, \infty)$. The impact of a rise in τ on the growth rate of private consumption is obtained from (2.16) as

$$\frac{\partial}{\partial \tau} \left(\frac{\dot{C}(t = 0, \tau)}{C(t = 0, \tau)} \right) = -h^{1-\alpha} \alpha (1 - \tau_p \varphi) < 0,$$

where again K and H are predetermined variables implying $\partial h(t = 0, \tau)/\partial \tau = 0$. This shows that at $t = 0$ the growth rate of private

---

[3] The economy in steady state means the same as the economy on the BGP.

consumption reduces as a result of the increase in the income tax rate τ and then rises gradually (since h rises) as the new BGP is approached. The same must hold for the private capital stock because we know from the foregoing that the growth rate of the private capital stock is smaller than that of consumption on the transition path. The impact of a rise in τ on the growth rate of public capital is obtained from (2.18) as

$$\frac{\partial}{\partial \tau} \left(\frac{\dot{H}(t=0,\tau)}{H(t=0,\tau)} \right) = h^{-\alpha}(1 - \tau_p \varphi) > 0,$$

where again h dose not change at $t = 0$. This result states that the growth rate of public capital rises and then declines over time (since h rises) as the new BGP is approached. This result was to be expected because an increase in the income tax rate at a certain point in time means that the instantaneous tax revenue rises. Because a certain part of the additional tax revenue is spent for public investment, the growth rate of public capital rises.

We summarize the results of our considerations in the following proposition.

Proposition 5 *Assume that the economy is on the BGP. Then a rise in the income tax rate leads to a temporary decrease in the growth rates of consumption and private capital but a temporary increase in the growth rate of public capital. Further, on the transition path the growth rates of public capital and consumption exceed the growth rate of private capital.*

Next, we analyze the effects of a rise in the pollution tax rate τ_p. To do so we proceed analogously to the case of the income tax rate. Doing the analysis it turns out that we have to distinguish between two cases. If $1 - \eta - \tau > 0$ the results are equivalent to those we derived for an increase in the income tax rate. If $1 - \eta - \tau < 0$ two different scenarios are possible.[4] First, if the new $h^\star$, that is, $h^\star$ after the increase in τ_p, is smaller than $-(1 - \eta - \tau)/(1 - \tau)$, the long-run values $h^\star$ and $c^\star$ decline. This holds because the new $\dot{h} = 0$ isocline lies above the old $\dot{h} = 0$ isocline, that is, the isocline before the increase in τ_p, for $h < -(1 - \eta - \tau)/(1 - \tau)$ and the $\dot{c} = 0$ isocline turns right with $c = \rho$ for $h = 0$ remaining unchanged. K and H are predetermined values, so the level of consumption must decrease and jump to the stable manifold, implying $\partial c(t = 0, \tau)/\partial \tau < 0$ to reach the new steady state $(c^\star, h^\star)$. Figure 3.2 shows the phase diagram.

---

[4] For $1 - \eta - \tau = 0$ the analysis is equivalent to that of a rise in the income tax rate with the only difference that $\partial(\dot{H}(t = 0, \tau_p)/H(t = 0, \tau_p))/\partial \tau_p = 0$ holds. Note that in this case the balanced growth rate declines.

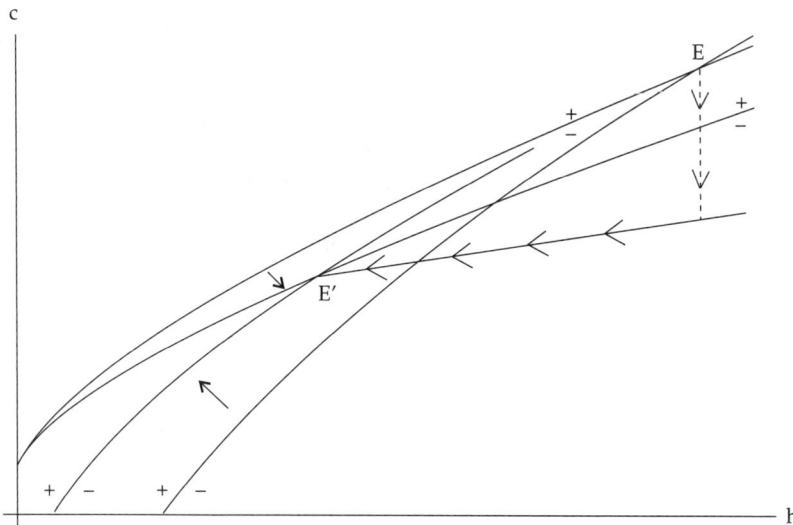

Figure 3.2 Effect of an increase in the pollution tax rate.

Over time both c and h decline until the new BGP is reached at (c^*, h^*). That is, we get $\dot{c}/c = \dot{C}/C - \dot{K}/K < 0$ and $\dot{h}/h = \dot{H}/H - \dot{K}/K < 0$, implying that on the transition path the growth rates of consumption and public capital are smaller than that of private capital for all $t \in [0, \infty)$. The impact of a rise in τ_p on the growth rate of private consumption is obtained from (2.16) as

$$\frac{\partial}{\partial \tau_p}\left(\frac{\dot{C}(t = 0, \tau_p)}{C(t = 0, \tau_p)}\right) = -h^{1-\alpha}\alpha(1 - \tau)\varphi < 0,$$

where again K and H are predetermined variables implying $\partial h(t = 0, \tau)/\partial \tau = 0$. This shows that at $t = 0$ the growth rate of private consumption falls as a result of the increase in the tax rate τ_p and then continues to decline gradually (since h declines) as the new BGP is approached. The impact of a rise in τ_p on the growth rate of public capital is obtained from (2.18) as

$$\frac{\partial}{\partial \tau_p}\left(\frac{\dot{H}(t = 0, \tau_p)}{H(t = 0, \tau_p)}\right) = h^{-\alpha}(1 - \eta - \tau)\varphi < 0, \quad \text{for } 1 - \eta - \tau < 0,$$

where again h does not change at $t = 0$. This result states that the growth rate of public capital declines and then rises over time (since h declines) as the new BGP is approached. As with a rise of the income tax rate, a higher pollution tax rate implies an instantaneous increase of the tax revenue. However, if η is relatively large, so that $1 - \eta - \tau < 0$, a large part of the additional tax revenue is used for abatement activities so that the growth rate of public capital declines although the tax revenue rises. The growth rate of the private capital stock may rise or decline. What we can say as to the the the growth rate of the private capital stock on the transition path is that it is always larger than those of consumption and of public capital.

Second, if the new $h^\star$, that is, $h^\star$ after the increase in τ_p, is larger than $-(1 - \eta - \tau)/(1 - \tau)$ the value for $h^\star$ rises while $c^\star$ may rise or fall. This holds because for $h > -(1 - \eta - \tau)/(1 - \tau)$ the new $\dot{h} = 0$ isocline lies below the old $\dot{h} = 0$ isocline, that is, the isocline before the increase in τ_p. In this case, the phase diagram is the same as the one in figure 3.1 with the exception that the $\dot{h} = 0$ isoclines before and after the rise in the tax rate intersect at $h = -(1 - \eta - \tau)/(1 - \tau)$. Another difference to the effects of a rise in the income tax rate is that the growth rate of public capital at $t = 0$ declines. The rest of the analysis is analogous to that of a rise in the income tax rate. In particular, we have again $\dot{c}/c = \dot{C}/C - \dot{K}/K > 0$ and $\dot{h}/h = \dot{H}/H - \dot{K}/K > 0$.

We can summarize our results in the following proposition.

Proposition 6 *Assume that the economy is on the BGP. Then a rise in the pollution tax rate shows the same temporary effects concerning the growth rates of consumption, private capital, and public capital as a rise in the income tax rate if $1 - \eta - \tau > 0$. If $1 - \eta - \tau < 0$, two situations are feasible: first, $h^\star$ declines and the temporary growth rates of consumption and public capital decline while the growth rate of private capital may rise or fall. Further, the temporary growth rates of consumption and public capital are smaller than that of private capital. Second, $h^\star$ rises and the temporary growth rates of consumption, public capital, and private capital fall. Further, on the transition path the growth rates of public capital and of consumption exceed the growth rate of private capital.*

In the next section we analyze welfare effects of fiscal policy assuming that the economy is on the BGP.

3.3 WELFARE EFFECTS OF FISCAL POLICY ON THE BGP AND THE SOCIAL OPTIMUM

3.3.1 Welfare Effects

In analyzing welfare effects, we confine our considerations to the model on the BGP. That is we assume that the economy immediately jumps to the new BGP after a change in fiscal parameters. In particular, we are interested in the question of whether growth and welfare maximization are identical goals.

To derive the effects of fiscal policy on the BGP arising from increases in tax rates at $t = 0$, we first compute (2.1) on the BGP as

$$W(\cdot) \equiv \arg \max_{C(t)} \int_0^\infty e^{-\rho t}(\ln C(t) - \ln P_E(t))dt. \qquad (3.3)$$

Denoting again the balanced growth rate by g, (3.3) can be rewritten as

$$W(\cdot) = \rho^{-1}\Big(\ln C_0 + \beta g/\rho + \beta \ln \eta + \ln \tau_p - (1 - \beta) \ln \varphi$$
$$- (1 - \beta) \ln K_0^\alpha H_0^{1-\alpha}\Big), \qquad (3.4)$$

with $C_0 = C(0)$, $K_0 = K(0)$, and $H_0 = H(0)$. From (2.16) and (2.17) we get

$$g = \alpha(1 - \tau)(1 - \tau_p\varphi)h^{1-\alpha} - \rho \quad \text{and} \quad C_0 = K_0 \Big((1 - \tau)(1 - \tau_p\varphi)h_0^{1-\alpha} - g\Big).$$

Combining these two expressions leads to $C_0 = K_0(\rho + g(1 - \alpha))/\alpha$. Inserting C_0 in (3.4) W can be written as

$$W(\cdot) = \rho^{-1}\left(\ln(\rho/\alpha + g(1 - \alpha)/\alpha) + \beta g/\rho + \ln \tau_p + C_1\right), \qquad (3.5)$$

with C_1 a constant given by $C_1 = \ln K_0 + \beta \ln \eta - (1 - \beta) \ln \varphi - (1 - \beta) \ln(K_0^\alpha H_0^{1-\alpha})$. Equation(3.5) shows that welfare in the economy positively varies with the growth rate on the BGP, that is, the higher the growth rate the higher welfare. Differentiating (3.5) with respect to τ and τ_p yields

$$\frac{\partial W}{\partial \tau} = \frac{\partial g}{\partial \tau}\left(c_0 \frac{1-\alpha}{\alpha \cdot \rho} + \frac{\beta}{\rho^2}\right), \quad \frac{\partial W}{\partial \tau_p} = \frac{1}{\tau_p \rho} + \frac{\partial g}{\partial \tau_p}\left(c_0 \frac{1-\alpha}{\alpha \cdot \rho} + \frac{\beta}{\rho^2}\right). \qquad (3.6)$$

With the expressions in (3.6) we can summarize our results in the following proposition.

Proposition 7 *Assume that the economy is in steady state and that there exist interior growth-maximizing values for the income and pollution tax rates. Then the welfare-maximizing pollution tax rate is larger than the growth-maximizing rate and the welfare-maximizing income tax rate is equal to the growth-maximizing income tax rate.*

Proof: The fact that the growth-maximizing income tax rate also maximizes welfare follows immediately from (3.6). Because the pollution tax rate τ_p maximizes the balanced growth we have $\partial g/\partial \tau_p = 0$. But (3.6) shows that for $\partial g/\partial \tau_p = 0$, $\partial W/\partial \tau_p > 0$ holds. Thus, the proposition is proved.   □

Proposition 7 states that welfare maximization may be different from growth maximization. If the government sets the income tax rate, it can be assured that the rate maximizing economic growth also maximizes welfare if one neglects transition dynamics. However, if the government chooses the pollution tax rate, it has to set the rate higher than that value maximizing the balanced growth rate to achieve maximum welfare. The reason for this outcome is that the pollution tax rate exerts a direct positive welfare effect by reducing effective pollution, in contrast to the income tax rate. This is seen from (3.5), where the expression $\ln \tau_p$ appears explicitly but τ does not.

The same also holds for variation of the parameter φ, which determines the degree of pollution as a byproduct of aggregate production. If φ declines, meaning that production becomes cleaner, there is always a positive partial and direct welfare effect going along with that effect. Again, this can be seen from (3.5), where $\ln \varphi$ appears. The overall effect of a decline in φ, that is, of the introduction of a cleaner technology, consists of this partial welfare effect and of changes in the balanced growth rate.

3.3.2 The Social Optimum

Next we compute the social optimum and study how the tax rates must be set so that the competitive economy replicates the social optimum. To do so, we first formulate the optimization problem of the social planner. Taking into account that $P_E = \varphi Y/A$ holds the social planner solves

$$\max_{C,A,I_p} \int_0^\infty e^{-\rho t} \left(\ln C + \ln A - \alpha \ln K - (1-\alpha) \ln H - \ln \varphi \right) dt,$$

subject to

$$\dot{K} = K^\alpha H^{1-\alpha} - C - A - I_p, \quad \dot{H} = I_p \quad \text{and} \quad 0 \le I_p \le \bar{I}_p < \infty.$$

The maximum principle yields $C = A = \lambda_1^{-1}$ and $\lambda_1 = \lambda_2$, with λ_1 the costate variable or the shadow price of private capital and λ_2 the shadow price of public capital. These conditions state that at any moment in time, the level of consumption and the level of abatement must be equal. This is obvious because consumption and abatement have the same effects on the utility of the household, and $\dot{K}$ is linear in both C and A. The level of public investment I_p has to be set such that the shadow price of private capital and that of public capital are equal for all $t \in [0, \infty)$. Since $\dot{K}$ and $\dot{H}$ are linear in I_p, there is a so-called bang-bang solution, that is, $I_p = 0$ must hold for $\lambda_1 < \lambda_2$ and $I_p = \bar{I}_p$ must hold for $\lambda_1 > \lambda_2$. In a way this can be seen as a no-arbitrage condition.

The shadow prices evolve according to

$$\dot{\lambda}_1 = \rho\lambda_1 - \lambda_1\alpha K^{\alpha-1}H^{1-\alpha} + \alpha/K \tag{3.7}$$

$$\dot{\lambda}_2 = \rho\lambda_2 - \lambda_1(1-\alpha)K^{\alpha}H^{-\alpha} + (1-\alpha)/H \tag{3.8}$$

Together with the conditions given by the maximum principle, these conditions are also sufficient if the limiting transversality condition $\lim_{t\to\infty} e^{-\rho t}(\lambda_1 K + \lambda_2 H) = 0$ holds. Proposition 8 characterizes the social optimum and shows how the fiscal parameters τ, τ_p, and η must be set so that the competitive economy replicates the social optimum.

Proposition 8 *Assume that the social optimum does not lead to sustained growth. Then the competitive economy replicates the social optimum if τ, τ_p, and η are chosen such that*

$$\frac{C}{K} = \left(\frac{1-\alpha}{\alpha}\right)^{1-\alpha}\left((1-\tau)(1-\varphi\tau_p) - 1\right) + 2\rho,$$

$$\tau = \frac{\varphi\tau_p(\eta-1)}{1-\varphi\tau_p}, \quad \varphi\tau_p = 1 - \frac{\rho(1-\alpha)^{\alpha-1}}{\alpha(1-\tau)} \tag{3.9}$$

hold for all $t \in [0,\infty)$. If the social optimum leads to ongoing growth, the competitive economy replicates the social optimum if τ, τ_p, and η are chosen such that

$$\eta = \rho\left(\frac{1-\alpha}{\alpha}\right)^{\alpha}\frac{1}{\varphi\tau_p} + \frac{\tau}{\varphi\tau_p} - \frac{(1-\tau)((1-\varphi\tau_p)\alpha - \varphi\tau_p)}{\varphi\tau_p}$$

$$\frac{\rho}{\alpha} = \left(\frac{1-\alpha}{\alpha}\right)^{1-\alpha} - \left(\frac{1-\alpha}{\alpha}\right)^{1-\alpha}(1-\tau)(1-\varphi\tau_p)$$

$$\frac{C}{K} = \left(\frac{1-\alpha}{\alpha}\right)^{1-\alpha}\left((1-\tau)(1-\varphi\tau_p) - \alpha\right) + 2\rho \tag{3.10}$$

hold for all $t \in [0,\infty)$. Further, the ratios of endogenous variables in the social optimum are given by

$$\frac{C}{K} = \frac{A}{K} = \frac{\rho}{\alpha}, \quad \frac{H}{K} = \frac{1-\alpha}{\alpha}, \quad \frac{I_p}{K} = (1-\alpha)\left(\left(\frac{1-\alpha}{\alpha}\right)^{1-\alpha} - 2\frac{\rho}{\alpha}\right). \tag{3.11}$$

Proof: Using $I_p = 0$ and setting equal the growth rates of K in the social optimum and the competitive economy yields the first equation in (3.9). The second equation is obtained by setting equal the growth rates of C

in the social optimum and the competitive economy. The third equation, finally, is obtained from $\dot{C}/C = 0$.

The first equation in (3.10) is obtained from setting $H/K = (1 - \alpha)/\alpha$ and then solving $f(\cdot) = 0$ from the proof of proposition 8 with respect to η. The second equation in (3.10) is obtained by equating the growth rate of consumption in the competitive economy with that in the social optimum.

The second equation in (3.11) is obtained by setting $\dot{\lambda}_1 = \dot{\lambda}_2$ and using $C = 1/\lambda_1$. This relation holds for both the BGP and for all $t \in [0, \infty)$. The first and third equations are obtained by using $H/K = (1 - \alpha)/\alpha$ and by setting $\dot{c}/c = 0$ and $\dot{h}/h = 0$, with $c = C/K$ and $h = H/K$.   □

Proposition 8 shows that $\eta > 1$ must hold so that the competitive economy can replicate the social optimum, in case the social optimum does not lead to ongoing growth. In this case, the model is the conventional neoclassical growth model with exogenous growth, which is well known and therefore was not considered explicitly in the previous section. The inequality $\eta > 1$ means that a certain part of the tax revenue must be used for abatement besides the revenue gained from taxing pollution. If the social optimum generates sustained growth, (3.10) shows how the fiscal parameters have to set such that the competitive economy replicates the social optimum.

The second equation in (3.11) states that the ratio of public to private capital equals the ratio of their elasticities in the social optimum. The third equation can also be written as $I_p/K = (1 - \alpha)(Y - C - A)/K$, implying $I/K = \alpha(Y - C - A)/K$, stating that the share of public to private investment is equal to the ratio of their elasticities with respect to output.

4

The Dynamics of the Model with Standard Preferences

In chapter 2 we presented and analyzed a growth model with environmental pollution and productive public capital. The analysis of the dynamics of the model demonstrated that it is characterized by local and global determinacy. However, we must also point out that this result may be due to the assumptions we made, especially concerning the utility function of the household.

Our goal in this chapter is to allow for a more general utility function and then give a complete characterization of the dynamics of our model as in Greiner (2007). We also intend to contribute to the literature on the dynamics of competitive economies with externalities.

Examples of such studies are the contributions by Benhabib and Farmer (1994) and by Benhabib et al. (2000). The difference of our work from these studies is twofold. First, we consider negative external effects of production, that is, pollution as a byproduct of production, in contrast to the aforementioned papers, which assume positive externalities associated with production or capital. Second, we do not assume that these externalities affect production in our economy but instead have negative repercussions on the utility of the household.

The structure of our economy is basically the same as in chapter 2, that is, we consider a decentralized economy consisting of three sectors: the household sector, a productive sector, and the government. Therefore, concerning the description of this model economy, we do not go into details but only point out differences compared to the previous chapter.

The representative household again maximizes its discounted stream of utility subject to its budget constraint:

$$\max_{C(t)} \int_0^\infty e^{-\rho t} V(t) dt, \tag{4.1}$$

with $V(t)$ the instantaneous subutility function as in the last section, which depends positively on the level of consumption, $C(t)$, and negatively on effective pollution, $P_E(t)$. However, in contrast to the last chapter, we do not assume that the utility function is additively separable in $C(t)$ and $P_E(t)$. Concretely, $V(t)$ now takes the following form

$$V(t) = (C(t)P_E(t)^{-\xi})^{1-\sigma}/(1-\sigma), \tag{4.2}$$

where $\xi > 0$ gives the disutility arising from effective pollution. $1/\sigma > 0$ is the intertemporal elasticity of substitution of private consumption between two points in time for a given level of effective pollution, and ln is the natural logarithm.

Setting the income tax rate equal to zero, $\tau = 0$, the budget constraint of the household is written as[1]

$$\dot{K} = -C + wL + rK. \tag{4.3}$$

Assuming that a solution to (4.1) subject to (4.3) exists, we can use the current-value Hamiltonian to describe that solution. The Hamiltonian function is written as

$$\mathcal{H}(\cdot) = (CP_E^{-\xi})^{1-\sigma}/(1-\sigma) + \lambda(-C + wL + rK), \tag{4.4}$$

with λ the costate variable. The necessary optimality conditions are given by

$$\lambda = C^{-\sigma}P_E^{-\xi(1-\sigma)}, \tag{4.5}$$

$$\dot{\lambda}/\lambda = \rho - r, \tag{4.6}$$

$$\dot{K} = -C + wL + rK. \tag{4.7}$$

Because the Hamiltonian is concave in C and K jointly, the necessary conditions are also sufficient if in addition the transversality condition at infinity $\lim_{t \to \infty} e^{-\rho t} \lambda(t) K(t) = 0$ is fulfilled. Moreover, strict concavity in C also guarantees that the solution is unique.

The productive sector in our economy is equivalent to the one in section 2.2. In particular, the production function is given by,

$$Y = K^\alpha L^{1-\alpha} H^{1-\alpha}. \tag{4.8}$$

Pollution is given by $P = \varphi Y$ and taxed at the rate $\tau_p > 0$. Firms take into account that one unit of output causes φ units of pollution for which they have to pay $\tau_p \varphi < 1$ per unit of output. Thus, the optimization problem of the firm is given by

$$\max_{K,L} K^\alpha L^{1-\alpha} H^{1-\alpha} (1 - \varphi \tau_p) - rK - wL. \tag{4.9}$$

---

[1] We again suppress the time argument if no ambiguity arises.

Assuming competitive markets and taking public capital as given, optimality conditions for a profit maximum are obtained as

$$w = (1 - \tau_p\varphi)(1 - \alpha)L^{-\alpha}K^{\alpha}H^{1-\alpha}, \tag{4.10}$$

$$r = (1 - \tau_p\varphi)\alpha K^{\alpha-1}H^{1-\alpha}L^{1-\alpha}. \tag{4.11}$$

The government in our economy is now assumed to receive tax revenue only from the taxation of pollution. Positing that the government again runs a balanced budget at any moment in time and setting the income tax rate equal to zero, the budget constraint of the government is written as[2]

$$I_p + A = \tau_p P \leftrightarrow I_p = \tau_p P(1 - \eta). \tag{4.12}$$

Abatement effects pollution in the same way as in 2.2, that is, $P_E = P/A^{\beta}$, and the evolution of public capital is described by

$$\dot{H} = I_p, \tag{4.13}$$

where for simplicity we again neglect depreciation of public capital.

In the following, labor is normalized to one, $L \equiv 1$. An equilibrium allocation in the economy, then, is given if $K(t)$ and $L(t)$ maximize profits of the firm, $C(t)$ maximizes (4.1), and the budget of the government is balanced.

Profit maximization of the firm implies that the marginal products of capital and labor equal the interest rate and the wage rate. This implies that in equilibrium the growth rate of physical capital is given by

$$\frac{\dot{K}}{K} = -\frac{C}{K} + \left(\frac{H}{K}\right)^{1-\alpha}(1 - \varphi\tau_p), \ K(0) = K_0. \tag{4.14}$$

Using the budget constraint of the government the growth rate of public capital is

$$\frac{\dot{H}}{H} = \left(\frac{H}{K}\right)^{-\alpha}\varphi\tau_p(1 - \eta), \ H(0) = H_0. \tag{4.15}$$

Utility maximization of the household yields the growth rate of consumption as

$$\frac{\dot{C}}{C} = -\frac{\rho}{\sigma} + \sigma^{-1}(1 - \varphi\tau_p)\alpha\left(\frac{H}{K}\right)^{1-\alpha} - \xi(1-\beta)\frac{1-\sigma}{\sigma}\left(\alpha\frac{\dot{K}}{K} + (1-\alpha)\frac{\dot{H}}{H}\right). \tag{4.16}$$

---

[2] The budget constraint is the same as in Bovenberg and de Mooij (1997), except that these authors also impose a tax on output.

Equations (4.14), (4.15), and (4.16) completely describe the economy in equilibrium. The initial conditions $K(0) = K_0$ and $H(0) = H_0$ are given and fixed, and $C(0)$ can be chosen freely by the economy. Further, the transversality condition $\lim_{t \to \infty} e^{-\rho t} \lambda(t) K(t) = 0$ must be fulfilled, with λ determined by (4.5).

Before we study the dynamics of the model, we recall the definition of a balanced growth path given in definition 1. According to that definition, all variables grow at the same rate on a BGP, implying that the ratios $c \equiv C/K$ and $h \equiv H/K$ are constant. Further, we defined a BGP as sustainable if the growth of instantaneous utility, $\dot{V}$, is strictly positive.

If the utility function is logarithmic in C and P_E, we saw that any BGP is also sustainable. However, for the more general utility function (4.2), this does not necessarily hold. From $\dot{V}/V = (1 - \sigma)(\dot{C}/C - \xi \dot{P}_E/P_E) = (1 - \sigma)(\dot{C}/C - \xi(1 - \beta)\dot{Y}/Y)$ we see that $\xi(1 - \beta) < 1$ is a necessary and sufficient condition for sustainability of the BGP.[3] From an economic point of view, this means that a BGP is sustainable if abatement is very effective, that is, if β is high, or if effective pollution does not have a string impact on utility, that is, if ξ is small.

The differential equation system describing the dynamics around a BGP is now written as

$$\frac{\dot{c}}{c} = -\frac{\rho}{\sigma} + \frac{\alpha h^{1-\alpha}(1 - \varphi\tau_p)}{\sigma} - (1 - \alpha)\xi(1 - \beta)\frac{1 - \sigma}{\sigma}h^{-\alpha}\varphi\tau_p(1 - \eta)$$

$$+ \left(1 + \alpha\xi(1 - \beta)\frac{1 - \sigma}{\sigma}\right)(c - h^{1-\alpha}(1 - \tau_p\varphi)), \qquad (4.17)$$

$$\frac{\dot{h}}{h} = c - h^{1-\alpha}(1 - \varphi\tau_p) + h^{-\alpha}\varphi\tau_p(1 - \eta). \qquad (4.18)$$

Concerning a rest point of system (4.17) and (4.18), note that again we only consider interior solutions. That means that we exclude the economically meaningless stationary point $c = h = 0$ such that we can consider our system in the rates of growth.[4] As to the existence and stability of a BGP, we can state the following proposition.

Proposition 9 *If $1 + \xi(1 - \beta)(1 - \sigma)/\sigma \geq 0$ there exists a unique BGP that is a saddle point.*

Proof: To prove the proposition we first calculate $c^\star$ on a BGP obtained from $\dot{h}/h = 0$ as

$$c^\star = h^{1-\alpha}(1 - \varphi\tau_p) - h^{-\alpha}\varphi\tau_p(1 - \eta).$$

---

[3] Note that $(1 - \sigma)$ and V have the same sign so that $(1 - \sigma)V > 0$ holds.
[4] Note also that h is raised to a negative power in (4.17).

assumed. We do not do this but instead assume that c on a BGP is positive.[7]

Next we consider the case $1 + \xi(1 - \beta)(1 - \sigma)/\sigma < 0$. The next proposition gives the dynamics in this case.

Proposition 10 *If $1+\xi(1-\beta)(1-\sigma)/\sigma < 0$, ρ must be sufficiently large for the existence of two BGPs. The BGP yielding the lower growth rate is saddle point stable, and the BGP giving the higher growth rate is asymptotically stable.*

Proof: To prove the proposition we recall from the proof of proposition 9 that a $h^\star$ such that $f(\cdot) \equiv \dot{c}/c = 0$ holds gives a BGP.

For $(1 + \xi(1 - \beta)(1 - \sigma)/\sigma) < 0$ we have

$$\lim_{h \to 0} f(\cdot) = \infty \quad \text{and} \quad \lim_{h \to \infty} f(\cdot) = \infty \quad \text{and}$$

$$\partial f(\cdot)/\partial h > =< 0 \Leftrightarrow h > =< h_{min} \quad \text{and} \quad \lim_{h \to 0} \partial f(\cdot)/\partial h = -\infty,$$

$$\lim_{h \to \infty} \partial f(\cdot)/\partial h = 0,$$

with $h_{min} = (-1)\alpha\varphi\tau_p(1 - \eta)(1 - \alpha)^{-1}(1 + \xi(1 - \beta)(1 - \sigma)/\sigma)(1 - \varphi\tau_p)^{-1}$.

This implies that $f(h, \cdot)$ is strictly monotonic decreasing for $h < h_{min}$, reaches a minimum for $h = h_{min}$, and is strictly monotonic increasing for $h > h_{min}$. This implies that there exist two BGPs (two points of intersection with the horizontal axis) if $f(h, \cdot)$ crosses the horizontal axis. This is guaranteed if $f(h_{min}, \cdot) < 0$ holds. Inserting h_{min} in $f(\cdot)$ gives

$$f(h_{min}, \cdot) = -\rho/\sigma - h_{min}^{-\alpha}\varphi\tau_p(1 - \eta)\,(1 + \xi(1 - \beta)(1 - \sigma)/\sigma)$$
$$\times (1 + \alpha/(1 - \alpha)).$$

A sufficient condition for $f(h_{min}, \cdot) < 0$ is a large ρ.

To analyze stability, we note that the determinant of the Jacobian can be written as $\det J = -c^\star h^\star \partial f(\cdot)/\partial h$. This shows that the first intersection point of $f(\cdot)$ with the horizontal axis (smaller $h^\star$ and thus larger balanced growth rate, see [4.15]) cannot be a saddle point because $\partial f(\cdot)/\partial h < 0$ holds at this point. This point is asymptotically stable (only negative eigenvalues or eigenvalues with negative real parts) if the trace is negative, i.e. if $tr J < 0$ holds. The trace of the Jacobian can be calculated as $tr J = -const. + c^\star\alpha(1 + \xi(1 - \beta)(1 - \sigma)/\sigma)$, which is negative for $(1 + \xi(1 - \beta)(1 - \sigma)/\sigma) < 0$.

---

[7] What can be said is that h on the BGP, $h^\star$, must be larger than $\varphi\tau_p(1 - \eta)/(1 - \varphi\tau_p)$ for $c^\star$ to be positive.

The second intersection point of $f(\cdot)$ with the horizontal axis (lower h^* and thus higher balanced growth rate, see [4.15]) is a saddle point because $\partial f(\cdot)/\partial h > 0$ holds at this point, implying $\det J < 0$.   □

Proposition 10 states that two BGPs can be observed in our model depending on parameter values. In this case, the model is both globally and locally indeterminate. However, neither of the two BGPs are sustainable. This holds because for $1 + \xi(1 - \beta)(1 - \sigma)/\sigma < 0$, σ must be larger than 1. $\sigma > 1$ implies that $\xi(1 - \beta) > 1$ must hold so that the inequality $1 + \xi(1 - \beta)(1 - \sigma)/\sigma < 0 \leftrightarrow \xi(1 - \beta)(\sigma - 1)/\sigma > 1$ can be fulfilled. But $\xi(1 - \beta) > 1$ implies that the BGP is not sustainable. Thus, the economy grows over time, but the growth rate of instantaneous utility is negative.

5

Pollution as a Stock

In the last chapter we assumed that pollution as a flow variable affects the utility of the household. In this chapter, we study our model economy assuming that production leads to emissions that build up a stock of pollution that negatively affects utility. We focus on the question of whether ongoing growth is feasible with a constant or even declining stock of pollution, and we derive both growth effects in the long run depending on the different scenarios.

First we describe the structure of the decentralized model economy where we start with the behavior of the economic agents. Basically, the structure of the economy is equivalent to that of the last chapter. Nevertheless, for the sake of readability, we briefly repeat it here.

5.1 THE HOUSEHOLD SECTOR

Again, our economy is represented by one household. The goal of this household is to maximize a discounted stream of utility arising from consumption $C(t)$ over an infinite time horizon subject to its budget constraint. As to the utility function, we use the function introduced in chapter 4:

$$\max_{C(t)} \int_0^\infty e^{-\rho t} V dt, \quad \text{where } V = \frac{C^{1-\sigma} X^{-\xi(1-\sigma)}}{1-\sigma}, \tag{5.1}$$

with $1/\sigma > 0$ the intertemporal elasticity of substitution of consumption between two points in time and $\xi > 0$ giving the disutility of additional pollution. For $\sigma = 1$ the utility function is logarithmic.[1] X gives cumulated pollution, and ρ is again the rate of time preference.

The budget constraint is given by

$$\dot{K} = (w + rK)(1 - \tau) - C, \tag{5.2}$$

with τ the income tax rate. The wage rate is denoted by w. The labor supply L is constant, supplied inelastically, and we normalize $L \equiv 1$. r gives the return to per capita capital K. The budget constraint (5.2) states that the individual has to decide how much to consume and how

---

[1] We do not consider the case $\sigma = 1$ in this chapter.

much to save, thus increasing consumption possibilities in the future. The depreciation of physical capital is again assumed to equal zero.

To derive necessary conditions, we formulate the Hamiltonian function as $\mathcal{H}(\cdot) = C^{1-\sigma}X^{-\xi(1-\sigma)}/(1-\sigma) + \lambda(-C + (w+rK)(1-\tau))$, with λ the costate variable. The necessary optimality conditions are given by

$$\lambda = C^{-\sigma}X^{-\xi(1-\sigma)},\tag{5.3}$$

$$\dot{\lambda}/\lambda = \rho - r(1-\tau),\tag{5.4}$$

$$\dot{K} = -C + (w+rK)(1-\tau).\tag{5.5}$$

Because the Hamiltonian is concave in C and K jointly, the necessary conditions are also sufficient if in addition the transversality condition at infinity $\lim_{t\to\infty} e^{-\rho t}\lambda(t)K(t) \geq 0$ is fulfilled.

5.2 THE PRODUCTIVE SECTOR AND THE STOCK OF POLLUTION

The productive sector in our economy is represented by one firm that chooses inputs to maximize profits and behaves competitively. Concerning pollution or emissions, $P(t)$, we suppose as in the last chapter that it is a byproduct of aggregate production $Y(t)$. In particular, we again set $P(t) = \varphi Y(t)$, with $\varphi = const. > 0$.

The stock of pollution X, which affects utility of the household, evolves over time according to the following differential equation (see Brock and Taylor 2004)

$$\dot{X} = P - A - \delta_x X,\ 0 < \delta_x,\tag{5.6}$$

where A gives abatement that reduces pollution and δ_x gives the exponential rate at which the environment dissipates pollution. Further, we posit that $X \geq 0$ holds with $X = 0$, giving the pristine state of the environment.

Emissions are taxed at the rate $\tau_p > 0$ and the firm takes into account that one unit of output causes φ units of emissions for which it has to pay $\tau_p\varphi$ per unit of output. The per capita production function is given by,

$$Y = K^\alpha H^{1-\alpha}L^{1-\alpha} \equiv K^\alpha H^{1-\alpha},\tag{5.7}$$

with H denoting the stock of productive public capital and $\alpha \in (0,1)$ gives the per capita capital share. Recall that K denotes per capita capital and that L is normalized to one.

Assuming competitive markets and taking public capital as given, the first-order conditions for a profit maximum are obtained as

$$w = (1 - \tau_p \varphi)(1 - \alpha) K^\alpha H^{1-\alpha}, \tag{5.8}$$

$$r = (1 - \tau_p \varphi) \alpha K^{\alpha-1} H^{1-\alpha}. \tag{5.9}$$

5.3 THE GOVERNMENT

The government in our economy uses resources for abatement activities A, which reduce the stock of pollution. Abatement activities $A \geq 0$ are financed by the tax revenue coming from the tax on emissions, that is, $A = \eta \tau_p E$, with $0 < \eta$. This means that η determines that part of the emission tax revenue used for abatement. Consequently, $1 - \eta$ gives that part of the revenue used for public investment in the public capital stock I_p, $I_p \geq 0$. For $\eta > 1$, a certain part of the tax revenue resulting from the income taxation is used for abatement in addition to the tax revenue gained by taxing emissions.

The government in this economy runs a balanced budget at any moment in time. Thus, the budget constraint of the government is written as

$$I_p = \tau_p E(1 - \eta) + \tau(w + rK). \tag{5.10}$$

The evolution of public capital is described by

$$\dot{H} = I_p, \tag{5.11}$$

where for simplicity we again assume that there is no depreciation of public capital.

5.4 EQUILIBRIUM CONDITIONS AND THE BALANCED GROWTH PATH

Before we define a balanced growth path we define an equilibrium.

Definition 2 *An equilibrium is a sequence of variables $\{C(t), K(t), H(t), X(t)\}_{t=0}^\infty$ such that given a sequence of prices $\{w(t), r(t)\}_{t=0}^\infty$, the firm maximizes profits, the household solves (5.1) subject to (5.2), and the budget constraint of the government (5.10) is fulfilled.*

Thus, in equilibrium the dynamics of the model are completely described by the following differential equation system:

$$\frac{\dot{C}}{C} = -\frac{\rho}{\sigma} + \frac{\alpha(1 - \tau)(1 - \varphi\tau_p)}{\sigma} \left(\frac{H}{K}\right)^{1-\alpha} + \xi\left(\frac{1-\sigma}{\sigma}\right)\delta_x$$

$$- \xi\left(\frac{1-\sigma}{\sigma}\right)\varphi\left(\frac{H}{K}\right)^{1-\alpha}\left(\frac{K}{X}\right)(1 - \eta\tau_p), \tag{5.12}$$

$$\frac{\dot{K}}{K} = -\frac{C}{K} + \left(\frac{H}{K}\right)^{1-\alpha} (1-\tau)(1-\varphi\tau_p), \ K(0) > 0, \tag{5.13}$$

$$\frac{\dot{H}}{H} = \left(\frac{H}{K}\right)^{-\alpha} \left(\varphi\tau_p (1-\eta) + \tau(1-\varphi\tau_p)\right), \ H(0) > 0, \tag{5.14}$$

$$\frac{\dot{X}}{X} = \varphi \left(\frac{H}{K}\right)^{1-\alpha} \left(\frac{K}{X}\right) (1-\eta\tau_p) - \delta_x, \ X(0) > 0. \tag{5.15}$$

The initial conditions $K(0)$, $H(0)$, and $X(0)$ are given and fixed, and $C(0)$ can be chosen freely by the economy. It should also be noted that we assume $X(0) > 0$, meaning that the environment initially is polluted.

In the following we examine our model concerning existence and stability of a balanced growth path (BGP). To do so, we define a BGP in this chapter as follows.

Definition 3 *A balanced growth path (BGP) is a path such that the economy is in equilibrium and such that consumption, private capital, and public capital grow at the same strictly positive constant growth rate, that is, $\dot{C}/C = \dot{K}/K = \dot{H}/H = g, g > 0, g = constant, and either*

(i) *$\dot{X} = 0$ or*
(ii) *$\dot{X}/X = -\delta_x$ or*
(iii) *$\dot{X}/X = \dot{C}/C = \dot{K}/K = \dot{H}/H = g$.*

This definition shows that on a BGP the growth rates of consumption, private capital, and public capital are positive and constant over time. Concerning the state of the environment, we consider three different scenarios. Scenario i describes a situation where the environmental quality is constant over time. Scenario ii describes an economy with an improving environment, and scenario iii finally gives an economy where the state of the environment deteriorates over time. Note that in scenarios i and ii the ratio X/K asymptotically converges to zero because cumulated pollution, X, is constant or declines in these scenarios, respectively, while K rises. In scenario iii, X/K is constant because X grows at the same rate as K in that scenario.

It should be pointed out that scenarios i and scenario ii can certainly be considered as reflecting a sustainable BGP because the environment is constant or even improves. This does not necessarily hold for scenario iii, where the environment deteriorates at the same rate as economic variables grow. If one adopts the definition of sustainability as given by Byrne (1997) and presented in the last chapter, scenario iii could also be sustainable. This holds because according to that definition, a BGP is said to be sustainable if instantaneous utility grows over time, that is, if $\dot{V} > 0$ holds. Depending on the parameters in the model this may hold for our economy. However, most contributions in the literature define a sustainable BGP as a path on which the quality of the environment is constant. In any case, we also consider scenario iii because it may serve as an approximation to real-world economies, which grow over time, which goes along with a deterioration of the environmental quality.

To analyze the model further, we first perform a change of variables. Defining $c = C/K$, $h = H/K$, and $x = X/K$ and differentiating these variables with respect to time gives $\dot{c}/c = \dot{C}/C - \dot{K}/K$, $\dot{h}/h = \dot{H}/H - \dot{K}/K$, and $\dot{x}/x = \dot{X}/X - \dot{K}/K$. A rest point of this new system then corresponds to a BGP of our original model economy. The system describing the dynamics around a BGP is given by

$$\dot{c} = c\left(c - \frac{\rho}{\sigma} + \frac{\alpha(1-\varphi\tau_p)(1-\tau)}{\sigma}h^{1-\alpha} - (1-\tau)(1-\varphi\tau_p)h^{1-\alpha}\right)$$

$$+ c\left(\xi\left(\frac{1-\sigma}{\sigma}\right)\delta_x - \xi\left(\frac{1-\sigma}{\sigma}\right)\varphi\left(\frac{H}{K}\right)^{1-\alpha}(1-\eta\tau_p)x^{-1}\right), \quad (5.16)$$

$$\dot{h} = h\left(c + h^{-\alpha}\left(\varphi\tau_p(1-\eta) + \tau(1-\varphi\tau_p)\right) - (1-\tau)(1-\varphi\tau_p)h^{1-\alpha}\right), \quad (5.17)$$

$$\dot{x} = x\left(c + \varphi h^{1-\alpha}x^{-1}(1-\eta\tau_p) - \delta_x - (1-\tau)(1-\varphi\tau_p)h^{1-\alpha}\right). \quad (5.18)$$

Concerning a rest point of system (5.16)–(5.18), note that we only consider interior solutions with respect to c and h, that is, we exclude the economically meaningless stationary point $c^\star = h^\star = 0$.

Before we continue and present results as concerns the dynamics of our economy, we note that scenario i is modeled by setting $\eta = (\varphi - \delta_x x h^{\alpha-1})/(\tau_p\varphi)$, which gives $\dot{X} = 0$ for all $t \in [0, \infty)$. Scenario ii is obtained by setting $\eta = 1/\tau_p$, which yields $\dot{X}/X = -\delta_x$ for all $t \in [0, \infty)$.

5.5 THE DYNAMICS OF THE MODEL

In this section we analyze the dynamics of the model economy. Thus, we study the question of whether a BGP exists, whether it is unique, and whether it is stable. As to the uniqueness and stability of a BGP for scenario i and for scenario ii we can state the following proposition.

Proposition 11 *Assume that $(\varphi - \tau)/\varphi(1-\tau) < \tau_p < 1/\varphi$ holds. Then there exists a unique saddle point stable balanced growth path for scenario i and for scenario ii.*

Proof: To prove this proposition, we solve $\dot{c}/c = 0$ with respect to c leading to $c = \rho/\sigma - h^{1-\alpha}\alpha(1-\tau)(1-\tau_p\varphi)/\sigma + (1-\tau)(1-\tau_p\varphi)h^{1-\alpha} +$

$\xi(1 - \sigma)\varphi h^{1-\alpha}(1 - \eta\tau_p)x^{-1}/\sigma - \xi\delta_x(1 - \sigma)/\sigma$. Inserting the expression in $\dot{h}$ gives for scenario i,

$$\dot{h} = h(\rho/\sigma + h^{-\alpha}\left(\varphi(\tau_p - 1) + \tau(1 - \varphi\tau_p)\right) - (1 - \tau)(1 - \varphi\tau_p)\alpha h^{1-\alpha}/\sigma),$$

where we used $\eta = (1/\tau_p) - \delta_x x h^{\alpha-1}/(\tau_p\varphi)$ and $x = 0$, for $t \to \infty$, because on the BGP K grows faster than X in scenarios i and ii. An $h^\star$ such that $\dot{h} = 0$ gives a BGP for our model. For scenario ii we get

$$\dot{h} = h(\rho/\sigma + h^{-\alpha}\left(\varphi(\tau_p - 1) + \tau(1 - \varphi\tau_p)\right)$$
$$- (1 - \tau)(1 - \varphi\tau_p)\alpha h^{1-\alpha}/\sigma - \xi(1 - \sigma)\delta_x/\sigma),$$

where we used $\eta = 1/\tau_p$ and $x = 0$, for $t \to \infty$. It is easily seen that $\lim_{h\to 0}\dot{h}/h = +\infty$, for $\tau_p > (\varphi - \tau)/\varphi(1 - \tau)$, and $\lim_{h\to\infty}\dot{h}/h = -\infty$, for $\tau_p < 1/\varphi$. Further, given these assumptions, $\partial(\dot{h}/h)/\partial h < 0$ holds so that the existence of a unique BGP is shown.

   To show saddle point stability, we compute the Jacobian matrix for scenarios, i and ii, which is given by

$$J = \begin{bmatrix} c^\star & \partial\dot{c}/\partial h & 0 \\ h^\star & \partial\dot{h}/\partial h & 0 \\ 0 & 0 & \dot{X}/X - \dot{K}/K \end{bmatrix},$$

where we used $x^\star = 0$. Because X is constant in scenario i and declines in scenario ii while $\dot{K}/K$ is strictly positive on the BGP, we get the first eigenvalue of this matrix as $\mu_1 = \dot{X}/X - \dot{K}/K < 0$. Further, it is easily seen that for scenarios i and ii, $c^\star(\partial\dot{h}/\partial h) - h^\star(\partial\dot{c}/\partial h) < 0$ holds so that the second eigenvalue is negative, too, and the third eigenvalue is positive.   □

   The assumption $(\varphi - \tau)/\varphi(1 - \tau) < \tau_p$ is equivalent to $\varphi\tau_p(1 - \eta) + \tau(1 - \varphi\tau_p) > 0$ in scenarios i and ii and is needed for $\dot{H}/H > 0$. It simply states that public investment must be positive for sustained growth, which is obvious because productive public spending is the source of ongoing growth in this model. Note that $(\varphi - \tau)/\varphi(1 - \tau) < 1/\varphi$ implies $\varphi < 1$. This means that scenarios i and ii are only feasible if an additional unit of output leads to less than one additional unit of pollution. Loosely speaking, this means that the technology in use must not be too polluting. This is intuitively plausible because it is difficult to have a constant or even improving quality of the environment with ongoing growth if production leads to a strong degradation of the environment. If production is very polluting, that is, $\varphi \geq 1$, a constant quality of the environment can be achieved only without sustained growth.

   Given the assumptions, proposition 11 shows that there exists a unique saddle point stable BGP for our model economy for scenarios i and ii. It should be mentioned that saddle point stability means that two eigenvalues of the Jacobian matrix are negative and one is positive. This implies that there exists a unique value for initial consumption $C(0)$ lying on the stable manifold of the rest point of (5.16)–(5.18). From an economic point of view, this result tells us that both global and local indeterminacy cannot arise in this economy with a constant or improving environment, that is, for scenarios i and ii.

For scenario iii, the situation is more complex. Chapter 4 has already shown there may be more than one BGP. Proposition 12 gives the exact outcome.

Proposition 12 *Assume that $\varphi\tau_p < 1$ and $\varphi\tau_p(1-\eta) + \tau(1-\varphi\tau_p) > 0$ hold in scenario iii. Then there exists a unique balanced growth path for $1/\sigma > (\xi - 1)/\xi$. For $1/\sigma < (\xi - 1)/\xi$, there exist either two BGPs or no BGP.*

Proof: To prove this proposition we set $\dot{h}/h = 0$, giving $c - (1 - \tau)$ $(1 - \tau_p\varphi) = -h^{-\alpha}(\varphi\tau_p(1-\eta) + \tau(1-\varphi\tau_p))$. Inserting this in $\dot{x}/x = 0$ leads to $\varphi h^{1-\alpha}(1 - \eta\tau_p)x^{-1} = \delta_x + h^{-\alpha}(\varphi\tau_p(1-\eta) + \tau(1-\varphi\tau_p))$. Using these relationships we get for $\dot{c}/c$,

$$\dot{c}/c = -\rho/\sigma + (1-\tau)(1 - \tau_p\varphi)\alpha h^{1-\alpha}/\sigma - h^{-\alpha}(\varphi\tau_p(1-\eta)$$
$$+ \tau(1 - \varphi\tau_p))(1 + \xi(1-\sigma)/\sigma).$$

For $1 + \xi(1-\sigma)/\sigma > 0 \leftrightarrow 1/\sigma > (\xi - 1)/\xi$, we get

$$\lim_{h\to 0}(\dot{c}/c) = -\infty, \quad \lim_{h\to\infty}(\dot{c}/c) = +\infty \quad \text{and} \quad \partial(\dot{c}/c)/\partial h > 0,$$

where $\varphi\tau_p < 1$ and $\varphi\tau_p(1-\eta) + \tau(1-\varphi\tau_p) > 0$ must hold. This shows that there exists a unique BGP for $1/\sigma > (\xi - 1)/\xi$.

If $1 + \xi(1-\sigma)/\sigma < 0 \leftrightarrow 1/\sigma < (\xi - 1)/\xi$ holds we get

$$\lim_{h\to 0}(\dot{c}/c) = +\infty \quad \text{and} \quad \lim_{h\to\infty}(\dot{c}/c) = +\infty$$

and

$$\frac{\partial(\dot{c}/c)}{\partial h} = \alpha(1-\alpha)h^{-\alpha}(1-\tau)(1-\tau_p\varphi)/\sigma + \alpha h^{-\alpha-1}$$
$$\times (\varphi\tau_p(1-\eta) + \tau(1-\varphi\tau_p))(1 + \xi\left(\frac{1-\sigma}{\sigma}\right),$$

with $\partial(\dot{c}/c)/\partial h \to -\infty$ (0), for $h \to 0$ (∞). For $h < (>)h_1$ the derivative is negative (positive), with $h_1 = -(\varphi\tau_p(1-\eta) + \tau(1-\varphi\tau_p))(1 + \xi$ $(1 - \sigma)/\sigma)/((1-\alpha)(1-\tau)(1-\tau_p\varphi)/\sigma) > 0$. Furthermore, there exists a unique inflection point of the function $\dot{c}/c$, h_2, given by $h_2 = $

$(1 + \alpha)h_1/\alpha > h_1$. This shows that, for $1/\sigma < (\xi - 1)/\xi$, there are either two points of intersection of $\dot{c}/c$ with the horizontal axis, and thus two BGPs or none.[2]   □

Proposition 12 demonstrates that the BGP is unique if the intertemporal elasticity of substitution is larger than one minus the inverse of the parameter determining the (dis)utility of additional pollution, that is, for $1/\sigma > (\xi - 1)/\xi$. From a qualitative point of view, this result is the same as in chapter 4, where a low intertemporal elasticity of substitution was a necessary condition for multiple BGPs, too. Note also that for a small effect of additional pollution on utility, that is, for $\xi \leq 1$, uniqueness of the BGP is always given, independent of the intertemporal elasticity of substitution. Thus we can state that global indeterminacy can only arise when the effect of pollution on utility is strong.

Concerning the dynamics around the BGP, this question is more difficult to answer. Proposition 13 gives insight into the local dynamics around a BGP for scenario iii.

Proposition 13 *Assume that there exists a unique BGP for scenario iii. Then the BGP is either saddle point stable or unstable. Assume that there exist two BGPs. Then the BGP yielding the lower growth rate is either saddle point stable or unstable, whereas the BGP yielding the higher growth rate is asymptotically stable or unstable.*

Proof: To prove this theorem, we compute the Jacobian matrix, which is obtained as

$$J = \begin{bmatrix} c^\star & \partial\dot{c}/\partial h & c^\star(x^\star)^{-2}(h^\star)^{1-\alpha}\varphi\xi(1-\sigma)(1-\eta\tau_p)/\sigma \\ h^\star & \partial\dot{h}/\partial h & 0 \\ x^\star & \partial\dot{x}/\partial h & -\varphi(h^\star)^{1-\alpha}(1-\eta\tau_p) \end{bmatrix}.$$

The sign of the determinant of the Jacobian is equivalent to the sign of $\alpha h^{-\alpha-1}(\varphi\tau_p(1-\eta) + \tau(1 - \varphi\tau_p))(1 + \xi(1-\sigma)/\sigma) + \alpha(1-\alpha)h^{-\alpha}(1-\tau)$ $(1 - \tau_p\varphi)/\sigma$. If the BGP is unique, we have $(1 + \xi(1-\sigma)/\sigma) > 0$, implying that $\det J > 0$ holds.

Denoting by $\mu_j, j = 1, 2, 3$, the jth eigenvalue of the Jacobian, we know that $\mu_1 \cdot \mu_2 \cdot \mu_3 = \det J$. Therefore, a positive determinant implies that there are either two negative eigenvalues (or two complex conjugate eigenvalues with negative real parts) or three positive eigenvalues (or one positive eigenvalue and two complex conjugate eigenvalues with positive real parts).

---

[2] We neglect the case where $\dot{c}/c$ is tangent to the horizontal axis, which has Lebesque measure zero.

If there are two BGPs, the BGP associated with the smaller $h^\star$ yields the higher growth rate and vice versa. From the proof of proposition 12 we know that $\dot{c}/c$ first intersects the horizontal axis from above and then from below. So at the first intersection point we have $\partial(\dot{c}/c)/\partial h < 0$, and at the second we have $\partial(\dot{c}/c)/\partial h > 0$. Because $\partial(\dot{c}/c)/\partial h = \alpha(1-\alpha)h^{-\alpha}(1-\tau)(1-\tau_p\varphi)/\sigma + \alpha h^{-\alpha-1}(\varphi\tau_p(1-\eta) + \tau(1-\varphi\tau_p))(1+\xi(1-\sigma/\sigma)$, we know that $\det J < 0$ at the smaller $h^\star$ and $\det J > 0$ at the higher $h^\star$. Thus, the BGP yielding the higher growth rate has either three negative eigenvalues (or one negative eigenvalue and two complex conjugate eigenvalues with negative real parts) or two positive eigenvalues (or two complex conjugate eigenvalues with positive real parts) and one negative. If $\det J > 0$ we have the same situation as for the case of a unique BGP. $\qquad\square$

Proposition 13 demonstrates that for scenario iii the dynamics may be more complex. For this scenario, the economy may be characterized by both global and local indeterminacy. We do not pursue this question further because it was extensively studied in the last chapter. In the next section we compare the different scenarios concerning the long-run growth rate.

5.6 EFFECTS OF THE DIFFERENT SCENARIOS ON THE BALANCED GROWTH RATE

In this section we study the question of which scenario generates a higher balanced growth rate. To do so we take scenario i, where environmental quality is constant over time, as a benchmark. Proposition 14 compares scenario i with scenario ii, which is characterized by an improving environmental quality.

Proposition 14 *For $1/\sigma > (<) 1$ scenario i is characterized by a lower (higher) balanced growth rate than scenario ii.*

Proof: To prove this proposition, we note from the proof of proposition 11 that a BGP in scenario i is given for a $h^\star$ such that

$$\dot{h} = h(\rho/\sigma + h^{-\alpha}(\varphi(\tau_p - 1) + \tau(1 - \varphi\tau_p)) - (1-\tau)(1-\varphi\tau_p)\alpha h^{1-\alpha}/\sigma) = 0$$

holds.

A BGP in scenario ii is given for an $h^\star$ such that

$$\dot{h} = h(\rho/\sigma + h^{-\alpha}(\varphi(\tau_p - 1) + \tau(1 - \varphi\tau_p)) - (1-\tau)(1-\varphi\tau_p)\alpha h^{1-\alpha}/\sigma)$$
$$- \xi\delta_x(1-\sigma)/\sigma = 0$$

holds.

This shows that the graph of $\dot{h}$ in scenario i is above (below) the graph of $\dot{h}$ in scenario ii for $1/\sigma > (<)1$, implying that scenario i gives a higher (lower) value h^*. The balanced growth rate is given by equation (5.14). Because of $x^* = 0$, $\eta = 1/\tau_p$ holds both in scenarios i and ii on the BGP, leading to the result in proposition 14.   □

Proposition 14 states that for a relatively high intertemporal elasticity of substitution, that is, for $1/\sigma > 1$, the scenario in which environmental quality improves, scenario ii, yields a higher balanced growth rate than the scenario where the environmental quality is constant, scenario i. This holds because for an intertemporal elasticity of substitution larger than one, the marginal utility of consumption is higher the smaller the stock of pollution is, that is, the better the environmental quality, and vice versa. This is seen by computing the cross derivative of the utility function which is given by

$$\frac{\partial^2 V}{\partial C \partial X} = -\xi(1-\sigma)C^{-\sigma}X^{-\xi(1-\sigma)-1} > (<)\,0 \leftrightarrow 1/\sigma < (>)\,1. \qquad (5.19)$$

Because in scenario ii the environmental quality is better than in scenario i, because the level of X declines in scenario ii while it is constant in scenario i, the household forgoes more consumption today and shifts it into the future in scenario ii compared to scenario i. Therefore, scenario ii implies a higher balanced growth rate for $1/\sigma > 1$.

Equation (5.19) suggests that consumption and a clean environment, that is, a small value of pollution, are complementary for $1/\sigma > 1$ because, in this case, marginal utility of consumption rises with a decline in the level of pollution. This means that marginal utility of consumption is higher the cleaner the environment is. For $1/\sigma < 1$, consumption and pollution can be considered substitutes because the marginal (dis)utility of additional pollution declines with a rising level of consumption. Thus, one could interpret proposition 14 such that scenario i leads to lower growth than scenario ii if consumption is complementary to a clean environment. If the household considers consumption as a substitute for the negative effect of pollution, scenario i generates a higher balanced growth rate than scenario ii.

Comparing scenario i with scenario iii, the analysis is more complex. Proposition 15 gives results as concerns the ratio of public capital to private capital on the BGP in these scenarios as well as concerning the long-run growth rate.

Proposition 15 *A necessary condition for scenario i to be associated with a higher h^* than scenario iii is $1/\sigma < 1$. A sufficient condition for scenario i to be associated with a lower h^* than scenario iii is $1/\sigma > 1$. Further, $1/\sigma > 1$ is a necessary condition for scenario i to generate a higher balanced growth rate than scenario iii.*

Proof: To prove this proposition we note from the proof of proposition 11 that an h^* such that

$$\dot{h}/h = \rho/\sigma - (1 - \tau)(1 - \varphi\tau_p)\alpha h^{1-\alpha}/\sigma$$

$$+ h^{-\alpha}\left(\varphi\tau_p(1 - \tau_p^{-1}) + \tau(1 - \varphi\tau_p)\right) = 0$$

gives a BGP for scenario i.

From the proof of proposition 12 we know that an h^* such that

$$-\dot{c}/c = \rho/\sigma - (1 - \tau)(1 - \tau_p\varphi)\alpha h^{1-\alpha}/\sigma + h^{-\alpha}\left(\varphi\tau_p(1 - \eta)\right.$$

$$+ \tau(1 - \varphi\tau_p)\left.\right)(1 + \xi(1 - \sigma)/\sigma) = 0$$

gives a BGP for scenario iii.

Further, we know that in scenario iii, $\eta < 1/\tau_p$ holds on the BGP. Setting $\eta = (1/\tau_p) - \epsilon$, with $\epsilon > 0$, we can write $-\dot{c}/c$ as

$$-\dot{c}/c = \rho/\sigma + h^{-\alpha}\left(\varphi\tau_p(1 - \tau_p^{-1}) + \tau(1 - \varphi\tau_p)\right)(1 + \xi(1 - \sigma)/\sigma)$$

$$+ h^{-\alpha}\varphi\tau_p\epsilon(1 + \xi(1 - \sigma)/\sigma) - (1 - \tau)(1 - \tau_p\varphi)\alpha h^{1-\alpha}/\sigma.$$

This shows that $1/\sigma > (<) 1$ is a sufficient (necessary) condition for the graph of $\dot{h}/h$ to lie above (below) the graph of $-\dot{c}/c$ so that h^* in scenario i is lower (higher) compared to scenario iii.

To derive growth effects, we first note that the balanced growth rate for scenario i, $g_{(i)}$, and for scenario iii, $g_{(iii)}$, are obtained from (5.14) as

$$g_{(i)} = (h^*_{(i)})^{-\alpha}\left(\varphi\tau_p(1 - \tau_p^{-1}) + \tau(1 - \varphi\tau_p)\right),$$

$$g_{(iii)} = (h^*_{(iii)})^{-\alpha}\left(\varphi\tau_p(1 - \tau_p^{-1}) + \tau(1 - \varphi\tau_p)\right) + (h^*_{(iii)})^{-\alpha}\varphi\tau_p\epsilon.$$

This shows that $h^*_{(i)} > (<) h^*_{(iii)}$ is sufficient (necessary) for a higher (lower) balanced growth rate in scenario iii compared to scenario i. Thus proposition 15 is proved. □

This proposition states that a high intertemporal elasticity of substitution is a necessary condition such that the balanced growth rate in scenario i, where environmental quality is constant, exceeds the growth rate of scenario iii, in which the state of the environment degrades over time. The mechanism behind this result is the same as in proposition 14. Thus, only with a large intertemporal elasticity of substitution, the household may be willing to forgo more consumption today and shift it into the future in scenario i compared to scenario iii, because the marginal utility of consumption rises with a cleaner environment when the intertemporal elasticity of substitution is larger than one.

Proposition 15 also shows that scenario i can never lead to a higher balanced growth rate than scenario iii if the ratio of public to private capital on the BGP in scenario iii is lower than in scenario i. Thus, for a small intertemporal elasticity of substitution, that is, for $1/\sigma < 1$, it is to be expected that scenario iii leads to higher long-run growth because $1/\sigma < 1$ is a necessary condition for a lower $h^\star$ in scenario iii compared to scenario i.

An additional result can be derived concerning the balanced growth rates in scenario iii and scenario i, depending on the number of BGPs in scenario iii. This holds because the number of BGPs in scenario iii depends on the relation between the intertemporal elasticity of substitution and the parameter giving the (dis)utility of pollution. This is the contents of the following corollary to proposition 15.

Corollary 1 *Assume that there exist two BGPs in scenario iii. Then scenario i yields a lower balanced growth rate than scenario iii.*

Proof: To prove this corollary, we recall from proposition 12 that the existence of two BGPs in scenario iii implies $1/\sigma < (\xi - 1)/\xi \leftrightarrow 1 + \xi(1-\sigma)/\sigma < 0$. The proof of proposition 15 shows that the latter inequality is sufficient for $h^\star$ in scenario i to be larger than in scenario iii so that the long-run growth rate in scenario iii is larger than that in scenario i.   □

When there exist two BGPs for scenario iii, the intertemporal elasticity of substitution of consumption is smaller than one minus the inverse of the parameter giving the (dis)utility of additional pollution. Then, the balanced growth rate in scenario iii goes along with a smaller value of public capital relative to private capital, $h^\star$, compared to scenario i so that the growth rate in scenario i is smaller than that of scenario iii. It should be noted that this holds for both BGPs in scenario iii. Thus, for a sufficiently small intertemporal elasticity of consumption, the balanced growth rates in the scenario with a declining environmental quality exceed both the balanced growth rate of the scenario with a constant environmental quality.

It should also be pointed out that $1/\sigma < 1 - 1/\xi$ is a sufficient condition for the balanced growth rates in scenario iii to exceed that of scenario i. Thus, corollary 1 gives a sufficient condition while the condition formulated in proposition 15, with respect to the intertemporal elasticity of substitution, is only necessary.

6

Concluding Remarks

In this part we presented an endogenous growth model with public capital and pollution. The main novelty of this approach compared to the literature on environmental pollution and endogenous growth is the assumption that pollution only affects the utility of the household and not production possibilities directly.

Analyzing our model, we derived the effects of fiscal policy on the long-run balanced growth rate. We demonstrated that variations in both the income tax rate and the pollution tax rate may have positive or negative growth effects. This holds because on one hand, the government finances productive public spending, which tends to raise economic growth, by the tax revenue, whereas, on the other hand, higher taxes reduce economic growth. Furthermore, we derived conditions that must be fulfilled so that an increase in these tax rates leads to a higher balanced growth rate. In particular, we could derive growth-maximizing values of tax rates without resorting to numerical simulations. Roughly speaking, a higher pollution tax rate can lead to higher growth only if the government spends a large part of its revenues for productive purposes.

In addition, we studied the effects of a rise in the income and pollution tax rate on the growth rates of consumption, private capital, and public capital on the transition path, and we have seen that the transition effects of fiscal policy may differ from the long-run effects.

We could also demonstrate that growth maximization and welfare maximization need not be equivalent goals on the balanced growth path. In particular, it turned out that the welfare-maximizing values of parameters are different from the growth-maximizing values if the parameters have a direct impact on effective pollution and, thus, on utility. For example, a pollution tax rate that exceeds its growth-maximizing value can generate higher welfare. Thus, lower growth and less environmental pollution can imply higher welfare compared to a situation characterized by high-output growth and more pollution. This seems to be of particular relevance for fast growing countries in Asia for example.

Finally, we derived the social optimum and showed how the government has to set fiscal parameters so that the competitive economy replicates the social optimum. Among other things, it was demonstrated that a certain part of the income tax revenue has to be spent for abatement besides the tax revenue gained from taxing pollution.

As to the dynamics of our growth model, we could demonstrate (again without resorting to numerical examples) that the model is both locally and globally determinate if the utility function is logarithmic in consumption. For the case of an isoelastic utility function, we could show that the parameters determining the negative effect of pollution on utility are crucial to the question of whether indeterminate equilibrium paths may exist. In general, it turned out that the stronger the external effect, the more likely are multiple balanced growth paths. Thus, environmental pollution can lead to indeterminacy.

In an extension of the model, we assumed that environmental pollution as a stock negatively affects utility of the household. In this variation we considered three different scenarios: first, we studied the scenario where the stock of pollution is constant; second, we analyzed the scenario with an improving environmental quality; and finally, we studied the scenario in which environmental pollution grows at the same positive rate as all other endogenous variables.

The analysis of this extension has demonstrated that the model is characterized by a unique saddle point stable balanced growth path for the scenarios with a constant or improving environmental quality. However, it also turned out that the latter scenarios are compatible with sustained economic growth only if the production technology in use is not too polluting. The model with the scenario where all endogenous variables grow at the same rate in the long run may also reveal a more complex dynamic outcome, just as in the previous chapter. Thus, both local and global indeterminacy may occur again.

Concerning the balanced growth rate, the analysis of the extended model has revealed that the intertemporal elasticity of substitution of consumption must be high such that the scenarios with a smaller stock of pollution are associated with a higher balanced growth rate. The reason for this outcome is that the household is more willing to forgo consumption and shift it into the future in scenarios with a cleaner environment, if the marginal utility of consumption in the future is higher, which is the only case if the intertemporal elasticity of substitution is relatively large.

PART II
Global Warming and
Economic Growth

7

Introduction and Overview

Another environmental problem, which has global dimensions, is global warming. The emission of greenhouse gases (GHGs), like carbon dioxide (CO_2) or methane (CH_4), has drastically increased in the twentieth century and continues to rise, leading to higher concentrations of GHGs in the atmosphere. Higher GHG concentrations can generate a rise in the average global surface temperature and may make extreme weather events more likely.

According to the Intergovernmental Panel on Climate Change (IPCC 2001, 2007), it is certain that the global average surface temperature of the Earth has increased since 1861. The global average surface temperature has risen by $0.6 \pm 0.2\,°C$ over the twentieth century. It is very likely[1] that the 1990s was the warmest decade and 1998 the warmest year since 1861 (IPCC 2001, p. 26) and the warming of the Earth continues. In addition, it is likely that statistically significant increases in heavy and extreme weather events have occurred in many mid- and high-latitude areas, primarily in the Northern Hemisphere.[2]

Changes in climate occur as a result of both internal variability within the climate system and external factors where the latter can be natural or anthropogenic. However, natural factors have made small contributions to the climate change observed over the past century. Instead, there is strong evidence that most of the warming observed over the past fifty years is the result of human activities.[3] Especially the emission of 6HGs are considered the cause for climate changes, and these emissions continue to alter the atmosphere in ways that are expected to affect the climate.

In the economics literature, numerous studies analyze the impact of environmental degradation on economic growth using endogenous growth models, as already mentioned in part I. These studies are rather simplified because they intend to derive general qualitative results showing the effects governmental measures can have. It is assumed that economic activities lead to environmental degradation and, as a consequence, reduce utility and production possibilities. The goal of these studies often is to analyze how public policy affects

---

[1] Very likely (likely) means that the level of confidence is between 90–99 (66–90) percent.
[2] More climate changes are documented in IPCC (2001, 2007), p. 34.
[3] For details on the role of human activities for global warming, see Stern (2006, 2007).

environmental conditions as well as the growth rate and welfare of economies. An example for such a model was presented in part 1.

However, studies that incorporate climate models in endogenous growth models and then study the effects of different time paths of GHG emissions on the growth rate of economies are less common. Instead, economic studies dealing with global warming are mostly cost-benefit analysis, which take the growth rate of economies as an exogenous variable. These studies then compute the discounted cost of reducing GHG emissions and confront them with the discounted benefit of a lower increase in GHG concentration and, as a consequence, a smaller increase in average global surface temperature. Thus, different environmental policies are evaluated.

Typically, the effect of global warming is modeled mostly using integrated assessment models. These are computable general equilibrium models in which stylized climatic interrelations are taken into account by a climate subsystem incorporated in the model. Examples for this type of models are CETA (see Peck and Teisberg 1992), FUND (see Tol 1999), RICE and DICE (see Nordhaus and Boyer 2000 and Nordhaus 2007b), WIAGEM (see Kemfert 2001), or DART (see Deke et al. 2001). Other examples for cost-benefit analyses are the contributions by Hackl and Pruckner (2003) and Tol (2001, 2003).[4] The goal of these studies, then, is to evaluate different abatement scenarios as to economic welfare and their effects on GHG emissions.[5] In analyzing implications of climate policies, these models often assume that the growth rate of the economy is exogenously given, and feedback effects of lower GHG concentrations in the atmosphere on economic growth are frequently neglected. For example, in Nordhaus and Boyer (2000) different abatement scenarios are analyzed where the growth rate of the economy is assumed to be an exogenous variable and the results are compared with the social optimum. In this study it is shown, among other results, that in all scenarios carbon taxes rise over time.

Another approach is taken by Uzawa (2003) who constructs a theoretical framework in which three major problems concerning global environmental issues are addressed. First, all aspects involved with global environmental issues exhibit externalities. Another aspect is that global warming involves international and intergenerational equity and justice. Finally, the problem of global warming concerns the management of the atmosphere, the oceans, and other natural resources that have to be decided by a consensus of all affected economies.

---

[4] We do not go into the details of these studies. The interested reader is referred to the IPCC report (1996) for the structure of such models.

[5] However, the results partly are very sensitive with respect to the assumptions made. See, for example, Popp (2003), who shows that the outcome in Nordhaus and Boyer (2000) changes when technical change is taken into account.

A great problem in studying the economic consequences of global warming is the uncertainty as to the damages caused by a change of the Earth's climate. Nevertheless, this is done. For example, the IPCC estimates that a doubling of CO_2, which goes along with an increase of global average surface temperature between 1.5 and 4.5 °C reduces world GDP by 1.5 to 2 percent (see IPCC 1996, p. 218). This damage is obtained for the economy in steady state and comprises both market and nonmarket impacts, where nonmarket impacts are direct reductions of people's welfare resulting from a climate change. But of course it must be repeated that there is great uncertainty in social cost estimates, especially concerning the direct impact of climate changes on individuals' welfare.

In this part, we study the interaction between global warming and economic growth. The model we use is basically in the line of the DICE94 framework presented by Nordhaus (1994, 2007b). The major extension of the framework consists in allowing for endogenous growth, implying that environmental policy does not only affect the level of economic variables but also the long-run growth rate. Because an important stylized fact of market economies is that sustained per capita growth can be observed without a tendency for declining growth rates, it seems necessary to incorporate that aspect in a model dealing with climate change and evaluate climate policies within this framework. Thus, we want to bring together models of endogenous growth and models dealing with changes in the climate on Earth.

First, we integrate a simple climate model in a descriptive model of endogenous growth to analyze the effects of GHG emissions and abatement policies on economic growth. We study a descriptive growth model because in a first step we want to focus on the interrelation between global warming, production activities, and economic growth leaving aside preferences. As far as the forces of growth that lead to sustained growth, we assume in chapter 9 that there are positive externalities of investment in physical capital that generate ongoing growth. Within this model we define the balanced growth path, which will be basically the same throughout this part, and we present some numerical examples to get an idea of how the economy reacts to different environmental policies.

In chapter 10 we present the *AK* endogenous growth model and compute the second-best abatement share as well as the socially optimal value for the investment share and the abatement share assuming a logarithmic utility function. The analysis in chapter 10 takes as a starting point the world as one homogeneous region and does not distinguish between different regions with different levels of emissions and different damages caused by global warming.

This shortcoming is taken into account in section 10.3, where different regions are assumed that differ with respect to their contributions to GHG emissions as well as the damages they suffer from a rise in temperature. This section, then, studies the situation for both the non-cooperative and cooperative cases. Further, the section analyzes the situation when the environmental instrument is set such that marginal damages are the same for all regions.

An economy where both the household sector and the productive sector optimize is presented in chapter 11. As to the forces of economic growth, we again adopt the approach with positive externalities of investment (see Greiner et al. 2005). This chapter analyzes the competitive economy and the social optimum and derives those values of fiscal parameters that make the competitive economy replicate the social optimum.

There is another important research direction, undertaken by scientists, that studies the impact of GHG emissions on climate change through the change of ocean circulations. The papers by Deutsch et al. (2002) and Keller et al. (2000), for example, describe how the Gulf Stream and the North Atlantic current, part of the North Atlantic thermohaline circulation (THC), transport a large amount of heat from warm regions to Europe. As those publications show, due to the heating of surface water, the currents could suddenly change and trigger a swift change in temperature. The THC collapse and the sudden cooling of regions would most likely have a strong economic impact on Europe and Africa. An event like this would have an impact on the climate in these regions and would also affect economic growth. Further results on THC mechanisms are given in Broecker (1997). In our modeling of the interaction of economic growth and climate change in 11.3, we leave aside this possible event, although it might exacerbate some of the results.

The overall goal of section 11.3 is different from the cited studies. Our primary goal is not to evaluate different abatement policies as to their welfare effects nor modeling exacerbating events for global warming. We want to study, in the context of a simple endogenous growth model, the long-run effects of the interaction of global warming and economic growth, in particular the transitions dynamics that might occur with global warming. More specifically, we want to study the question of whether there possibly exist multiple equilibria and thresholds that separate basins of attraction for optimal paths to some long-run steady state. To study such a problem, we take the basic endogenous growth model from this chapter as the starting point and integrate a simple nonlinear feedback effect.

We also want to point out some limitations of the model. We do not intend to use an elaborate large-scale macro model describing the process of global warming. Instead, we confine our analysis to a basic energy balance model (EBM) that allows for feedback effects. As to those feedback effects we posit that the albedo of the Earth is affected by increases in GHGs. Other possible feedback effects, such as a change in the flux ratio (see chapter 8) for example, are neglected. We are aware that this limits the relevance of our model in a way. However, the qualitative outcome and the message of this book remain the same with our simplified specification.

Further, we also point out that we pursue an approach that assumes that losses due to temperature increases can be expressed in terms of GDP. This implies that the environment does not have an intrinsic value for humankind or that the loss can be substituted by consuming an additional amount of goods. This approach may be seen as problematic because the possible value of an unpolluted environment in the future may not yet be known. Furthermore, destroying large parts of the environment and species without even knowing of their existence may be extremely costly because its potential can never be exploited.

Taking these arguments seriously would require a different approach where the destruction of the environment equals its regenerative capacity so that the constancy of the environmental quality is a constraint that has to be fulfilled in solving an economic problem, like welfare maximization, for example. For the problem of global warming, this would mean that the GHG concentration remains at its preindustrial level, which is not a realistic assumption. So allowing for a framework where the GHGs exceed their preindustrial level seems to be more appropriate. Nevertheless, the problem of estimating future costs going along with environmental damages, whether caused by global warming or by other human activities, is difficult and should be addressed with care. More details concerning the problem of estimating costs associated with global warming are given in Azar and Schneider (2003), Gerlagh and Papyrakis (2003), and Weitzman (2007a,b). The latter in particular discusses the role of the discount rate in evaluating the cost of future damages.

8

Facts on GHG Emissions and the Change in Average Global Surface Temperature

Before we present the economic framework, we begin with a description of scientific knowledge concerning GHG emissions and the change in global average surface temperature. The simplest method of considering the climate system of the Earth is in terms of its global energy balance which is done by so-called energy balance models (EBM). According to an EBM, the change in the average surface temperature on Earth is described by[1]

$$\frac{dT(t)}{dt} c_h \equiv \dot{T}(t) c_h = S_E - H(t) - F_N(t), \ T(0) = T_0, \qquad (8.1)$$

with $T(t)$ the average global surface temperature measured in Kelvin (K),[2] c_h the heat capacity[3] of the Earth with dimension $J\,m^{-2}\,K^{-1}$ (Joules per square meter per Kelvin),[4] which is considered a constant parameter, S_E is the solar input, $H(t)$ is the nonradiative energy flow, and $F_N(t) = F \uparrow (t) - F \downarrow (t)$ is the difference between the outgoing radiative flux and the incoming radiative flux. S_E, $H(t)$, and $F_N(t)$ have the dimension Watts per square meter (Wm^{-2}). t is the time argument, which will be omitted in the following as long as no ambiguity can arise. $F \uparrow$ follows Stefan-Boltzmann-Gesetz, which is

$$F \uparrow = \epsilon \, \sigma_T \, T^4, \qquad (8.2)$$

with ϵ the emissivity that gives the ratio of actual emission to blackbody emission. Blackbodies are objects that emit the maximum amount of radiation and have $\epsilon = 1$. For the Earth, ϵ can be set to $\epsilon = 0.95$. σ_T is the Stefan-Boltzmann constant, which is given by $\sigma_T = 5.67\,10^{-8}\,Wm^{-2}K^{-4}$.

---

[1] This part follows Roedel (2001) chapter 10.2.1 and chapter 1. See also Henderson Sellers and McGuffie (1987) and Gassmann (1992). A more complex presentation can be found in Harvey (2000).
[2] 273 Kelvin equals 0 °C.
[3] The heat capacity is the amount of heat that needs to be added per square meter of horizontal area to raise the surface temperature of the reservoir by 1K.
[4] 1 Watt is 1 Joule per second.

Further, the ratio $F\uparrow/F\downarrow$ is given by $F\uparrow/F\downarrow = 109/88$. The difference $S_E - H$ can be written as $S_E - H = Q(1 - \alpha_1)\alpha_2/4$, with $Q = 1367.5\,Wm^{-2}$ the solar constant, $\alpha_1 = 0.3$ the planetary albedo, determining how much of the incoming energy is reflected by the atmosphere, and α_2 ($\alpha_2 = 0.3$) captures the fact that a part of the energy is absorbed by the surface of the Earth.

Summarizing this discussion, the EBM can be rewritten as

$$\dot{T}(t)\,c_h = \frac{1367.5}{4}\,0.21 - 0.95\left(5.67\,10^{-8}\right)(21/109)\,T^4, \quad T(0) = T_0. \quad (8.3)$$

In equilibrium, that is, for $\dot{T} = 0$, (8.3) gives a surface temperature of about 288.4K which is about 15.4°C. c_h is the heat capacity of the Earth. Because most of the Earth's surface is covered by sea water, c_h is largely determined by the oceans. Therefore, the heat capacity of the oceans is used as a proxy for that of the Earth. c_h is then given by $c_h = \rho_w\,c_w\,d\,0.7$, with ρ_w the density of sea water ($1027\,m^{-3}\,kg$), c_w the specific heat of water ($4186\,J\,kg^{-1}\,K^{-1}$), and d the depth of the mixed layer, which is set to 70 m. The constant 0.7 results from the fact that 70 percent of the Earth is covered with sea water. Inserting the numerical values, assuming a depth of 70 m and dividing by the surface of the Earth gives $c_h = 0.1497\,J\,m^{-2}\,K^{-1}$.

The effect of emitting GHGs is to raise the concentration of them in the atmosphere, which increases the greenhouse effect of the Earth. This is done by calculating the so-called radiative forcing, which is a measure of the influence a GHG, like CO_2 or CH_4, has on changing the balance of incoming and outgoing energy in the Earth-atmosphere system. The dimension of the radiative forcing is Wm^{-2}. For example, for CO_2 the radiative forcing, which we denote as F, is given by

$$F = 6.3\ln\frac{M}{M_o}, \quad (8.4)$$

with M the actual CO_2 concentration, M_o the preindustrial CO_2 concentration, and ln the natural logarithm (see IPCC 1990, p. 52–53).[5] For other GHGs, other formulas can be given describing their respective radiative forcing, and these values can be converted in CO_2 equivalents. Incorporating (8.4) in (8.3) gives

$$\dot{T}(t)\,c_h = \frac{1367.5}{4}\,0.21 - 0.95\left(5.67\,10^{-8}\right)(21/109)T^4$$

$$+ \beta_1\,(1 - \xi)\,6.3\ln\frac{M}{M_o}, \quad T(0) = T_0. \quad (8.5)$$

---

[5] The CO_2 concentration is given in parts per million (ppm).

Table 8.1 Important Parameters Used in the EBM.

Parameter	Meaning	Adopted Value	Unit
c_h	Heat capacity	0.1497	$Jm^{-2}K^{-1}$
σ_T	Stefan-Boltzmann constant	$5.67\ 10^{-8}$	$Wm^{-2}K^{-1}$
Q	Solar constant	1367.5	Wm^{-2}
ξ	Part of temperature rise absorbed by oceans	0.23	percentage
β_1	Feedback effect	1.1	percentage
β_2	Emissions absorbed by oceans	0.49	percentage
μ	Inverse of atmospheric lifetime of GHG	0.1	percentage

β_1 is a feedback factor that captures the fact that a higher GHG concentration affects, for example, atmospheric water vapor, which has effects for the surface temperature on Earth. β_1 is assumed to take values between 1.1 and 3.4. The parameter ξ, finally, captures the fact that $\xi = 0.3$ of the warmth generated by the greenhouse effect is absorbed by the oceans, which transport the heat from upper layers to the deep sea. Setting $\beta_1 = 1.1$ and assuming a doubling of GHGs implies that in equilibrium the average surface temperature rises from 15.4 to 18.7°C, implying a rise of about 3.3 degrees. This is in the range of IPCC estimates,[6] which yield increases between 1.5 and 4.5°C as a consequence of doubling GHG concentration (IPCC 2001, p. 67).

The concentration of GHGs M evolves according to the following differential equation

$$\dot{M} = \beta_2 E - \mu M, M(0) = M_0. \tag{8.6}$$

E denotes emissions and μ is the inverse of the atmospheric lifetime of GHGs. As to the parameter μ, we assume a value of $\mu = 0.1$.[7] β_2 captures the fact that a certain part of GHG emissions are taken up by oceans and do not enter the atmosphere. According to IPCC, $\beta_2 = 0.49$ for the time period 1990 to 1999 for CO_2 emissions (IPCC 2001, p. 39).

Table 8.1 gives a survey of the parameters used in our energy balance model.

---

[6] IPCC results are obtained with more sophisticated atmosphere-ocean general circulation models.

[7] The range of μ for CO_2 given by IPCC is $\mu \in (0.005, 0.2)$, see IPCC (2001), p. 38.

9

A Descriptive Model of Endogenous Growth

9.1 STRUCTURE OF THE MODEL

We start with the description of our growth model. We assume that aggregate production takes place according to the following aggregate production function (see Greiner 2004a)

$$\bar{Y}(t) = A\bar{K}(t)^{\alpha}(H(t)L(t))^{1-\alpha}D(T(t) - T_o), \qquad (9.1)$$

with $\bar{Y}(t)$ aggregate production, A a positive constant, $H(t)$ human capital or a stock of knowledge formed as a byproduct of aggregate investment, and $L(t)$ labor input. $\bar{K}$ is aggregate physical capital, $\alpha \in (0,1)$ is the capital share, and t gives time. $D(T(t) - T_o)$ is the damage function giving the damage resulting from deviations of actual temperature from preindustrial temperature, T_o. It should be mentioned that the assumption of a continuous damage function is only justified provided the temperature increase does not exceed a certain threshold. This holds because for higher increases of the temperature, catastrophic events may occur, along with extremely high economic costs that cannot be evaluated. Just one example is the breakdown of the Gulf Stream, which would dramatically change the climate in Europe. Therefore, the analysis assuming a damage function only makes sense for temperature increases within certain bounds.

Per capita production is obtained by dividing both sides of (9.1) by L as

$$Y = AK^{\alpha}H^{1-\alpha}D(\cdot). \qquad (9.2)$$

The income identity in per capita variables in the economy is given by

$$Y - X = I + C + B, \qquad (9.3)$$

with $X = \tau Y$, $\tau \in (0,1)$, the (per capita) tax revenue, I investment, C consumption, and B abatement activities. This means that national income after tax is used for investment, consumption, and abatement. As for abatement activities, we assume that this variable is expressed as ratio to total tax revenue X,

$$B = \tau_b X = \tau_b \tau Y, \tag{9.4}$$

with $\tau_b \in (0,1)$ the ratio of abatement spending to the tax revenue.

As for the damage function $D(T - T_o)$, we assume that it is C^2 and satisfies

$$D(T - T_o) \begin{cases} = 1, & \text{for } T = T_o \\ < 1, & \text{for } T \neq T_o, \end{cases} \tag{9.5}$$

with derivative

$$\frac{\partial D(\cdot)}{\partial T} \equiv D'(\cdot) \begin{cases} > 0, & \text{for } T < T_o \\ < 0, & \text{for } T > T_o. \end{cases} \tag{9.6}$$

The per capita capital accumulation function is given by

$$\dot{K} = AK^\alpha H^{1-\alpha} D(\cdot)(1 - \tau) - C - B - (\delta + n)K$$
$$= AK^\alpha H^{1-\alpha} D(\cdot)(1 - \tau(1 + \tau_b) - c(1 - \tau)) - (\delta + n)K, \tag{9.7}$$

with $n \in (0,1)$ the growth rate of labor input and $\delta \in (0,1)$ is the depreciation rate of physical capital. Further, we express consumption as ratio to GDP after tax, that is, $C = cY(1 - \tau), c \in (0,1)$. It should be noted that for $\dot{K} > 0$ the parameters must be such that $\tau(1 + \tau_b) + c(1 - \tau) \in (0,1)$ holds.

As mentioned, we assume that gross investment in physical capital is associated with positive externalities that build up a stock of knowledge capital that positively affects labor productivity. Knowledge per capita evolves according to

$$\dot{H} = \varphi I - (\eta + n)H = \varphi(AK^\alpha H^{1-\alpha} D(\cdot)(1 - \tau) - C - B) - (\eta + n)H$$
$$= \varphi(AK^\alpha H^{1-\alpha} D(\cdot)(1 - \tau(1 + \tau_b) - c(1 - \tau))) - (\eta + n)H, \tag{9.8}$$

with $\varphi > 0$ a coefficient determining the external effect associated with investment and $\eta \in (0,1)$ depreciation of knowledge.

It should also be mentioned that our formulation implies that government spending except for abatement does not affect production possibilities. Concerning emissions of GHGs, we assume that these are a byproduct of production and expressed in CO_2 equivalents. So emissions are a function of per capita output relative to per capita abatement activities. This implies that a higher production goes along with higher emissions for a given level of abatement spending. This assumption is frequently encountered in environmental economics (see, e.g., Smulders 1995). It should also be mentioned that the emission of GHGs does not affect production directly but only indirectly by affecting the climate of the Earth, which leads to a higher surface temperature and to more extreme weather situations. Formally, emissions are described by

$$E = \left(\frac{aY}{B}\right)^{\gamma},$$ (9.9)

with $\gamma > 0$ and $a > 0$ constants. The parameter a can be interpreted as a technology index describing how polluting a given technology is. For large values of a, a given production (and abatement) goes along with high emissions, implying a relatively polluting technology and vice versa.

The economy is completely described by equations (9.7), (9.8), (8.5), and (8.6), with emissions given by (9.9).

9.2 THE BALANCED GROWTH PATH

The balanced growth path is defined in this chapter as follows[1]

Definition 4 *A balanced growth path (BGP) is a path such that $\dot{T} = 0$, $\dot{M} = 0$, and $\dot{K}/K = \dot{H}/H$ hold, with $M \geq M_0$.*

This definition contains several aspects. First, we require that the temperature and the GHG concentration must be constant along a BGP. This is a sustainability aspect. Second, the growth rate of per capita capital equals that of per capita knowledge and is constant. Note that this implies that the growth rates of per capita GDP and per capita consumption are constant, too, and equal to that of capital and knowledge. Third, we only consider BGPs with a GHG concentration larger than or equal to the preindustrial level. This requirement is made for reasons of realism. Because GHG concentration has been rising monotonically over the past decades, it is not necessary to consider a situation with declining GHG concentration. The next proposition shows that there exists a unique BGP for this economy.

Proposition 16 *For the model economy, there exists a unique BGP that is asymptotically stable.*

Proof: First, we define $k = K/H$, giving $\dot{k}/k = \dot{K}/K - \dot{H}/H$.[2] To show uniqueness of the steady state we solve $\dot{k}/k = 0$, (8.5) = 0, and (8.6) = 0

---

[1] In the following, steady state is used equivalently to balanced growth path.
[2] Since k is raised to a negative power in (9.7)/K, $k = 0$ is not feasible and we therefore consider the equation $\dot{k}/k$ in the rate of growth.

with respect to k, T, and M. Setting (8.5) $= 0$ gives

$$T_{1,2} = \pm 99.0775 \, (71.7935 + 4.851 \ln(M/M_0))^{1/4}$$

$$T_{3,4} = \pm 99.0775\sqrt{-1} \, (71.7935 + 4.851 \ln(M/M_0))^{1/4}.$$

Clearly $T_{3,4}$ are not feasible. Further, because $M \geq M_0$, only the positive solution of $T_{1,2}$ is feasible. Uniqueness of M is immediately seen. The equation $\dot{k}/k$ is given by

$$f(k, \cdot) \equiv \frac{\dot{k}}{k} = Ak^{\alpha-1}D(\cdot)(1 - \varphi k)((1 - \tau)(1 - c) - \tau\tau_b) - (\delta - \eta),$$

with $\partial f(k, \cdot)/\partial k < 0$ and $\lim_{k \to 0} f(k, \cdot) = +\infty$ and $\lim_{k \to \infty} f(k, \cdot) = -\infty$ for a given T. This shows that there exists a unique k that solves $\dot{k}/k = 0$.

To study the local dynamics, we calculate the Jacobian matrix J corresponding to this dynamic system, which is obtained as

$$J = \begin{pmatrix} \partial \dot{k}/\partial k & D'(\cdot)A(k^\star)^\alpha(1 - \varphi k^\star)((1 - \tau)(1 - c) - \tau\tau_b) & 0 \\ 0 & -(79.8 \, (5.67 \, 10^{-8}) \, (T^\star)^3)/(109c_h) & 4.851/(c_h M^\star) \\ 0 & 0 & -\mu \end{pmatrix},$$

with $\star$ denoting steady-state values and the parameter values as in section 9.3. The eigenvalues of J are given by

$$e_1 = -\mu, \quad e_2 = -(79.8 \left(5.67 \, 10^{-8} \right)(T^\star)^3)/(109c_h) \quad \text{and} \quad e_3 = \partial \dot{k}/\partial k.$$

Because $\partial f(k, \cdot)/\partial k < 0$, $\partial \dot{k}/\partial k < 0$ also holds. Thus, the proposition is proved. $\qquad\square$

This proposition shows that any solution starting in the vicinity of the BGP will converge to this path in the long run. The balanced growth rate of the economy is given by (9.7)/K as

$$g \equiv A(k^\star)^{\alpha-1}D(\cdot)((1 - \tau)(1 - c) - \tau\tau_b) - (\delta + n), \qquad (9.10)$$

with $k^\star$ the value of k on the BGP, where k is defined as $k \equiv K/H$. However, it cannot be excluded that the BGP goes along with a negative growth rate because the sign of the balanced growth rate depends on the concrete numerical values of the parameters. The question of whether there exists a BGP with a positive growth rate for a certain parameter constellation is addressed in the next section. In this section we make the assumption that a non-degenerate BGP exists and analytically study growth effects of varying the tax rate and abatement spending.

In the next step we analyze how the balanced growth rate reacts to changes in the income tax rate τ and to different values of the ratio τ_b. To do so we differentiate g with respect to τ. This gives

$$\frac{\partial g}{\partial \tau} = AD(\cdot)(k^\star)^{\alpha-1}(-1)(1 - c + \tau_b)$$

$$+ (\alpha - 1)(k^\star)^{\alpha-2}\frac{\partial k}{\partial \tau}AD(\cdot)((1 - \tau)(1 - c) - \tau\tau_b)$$

$$+ D'(\cdot)\frac{\partial T}{\partial \tau}(k^\star)^{\alpha-1}A((1 - \tau)(1 - c) - \tau\tau_b) > < 0 \qquad (9.11)$$

From (8.5) and (8.6) it is easily seen that $\partial k^\star/\partial \tau < 0$ and $\partial T^\star/\partial \tau < 0$, for $T > T_o$. To see this one uses that $T^\star$ positively depends on $M^\star$, which, for its part, negatively depends on τ. The latter is seen by calculating $M^\star$ from $\dot{M} = 0$ and using (9.9). $\partial k^\star/\partial \tau < 0$ is obtained from implicitly differentiating $\dot{k}/k = \dot{K}/K - \dot{H}/H$.

The second inequality in (9.11) results from the fact that in our model an increase in the tax revenue raises abatement activities since we assume a fixed ratio of abatement activities to tax revenue. The first inequality states that a rise in the tax rate reduces the ratio of physical to human capital. This shows that an increase in the tax rate has both positive and negative partial growth effects. On the one hand, a higher tax rate reduces investment because more resources are spent for abatement.

On the other hand, a higher tax rate reduces the increase in average global surface temperature and, as a consequence the damage resulting from an increase in T. This raises aggregate production which has a positive growth effect. Further, it should be mentioned that the decrease in $K/H = k$ has also a positive growth effect since a lower ratio of physical to human capital goes along with higher growth. This means that economies with small physical capital stocks and high stocks of knowledge are likely to show large growth rates. This, for example, was the case for Germany and Japan after the Second World War. So the analytical model does not answer the question of whether a higher tax rate reduces or increases the long-run balanced growth rate.

Next, we analyze the effects of an increase in abatement activities implying a higher value of τ_b. The derivative of g with respect to τ_b is given by

$$\frac{\partial g}{\partial \tau_b} = AD(\cdot)(k^\star)^{\alpha-1}(-\tau)$$

$$+ AD(\cdot)(\alpha - 1)(k^\star)^{\alpha-2}\frac{\partial k}{\partial \tau_b}((1 - \tau)(1 - c) - \tau\tau_b)$$

$$+ AD'(\cdot)\frac{\partial T}{\partial \tau_b}(k^\star)^{\alpha-1}((1 - \tau)(1 - c) - \tau\tau_b) > < 0. \qquad (9.12)$$

As for the tax rate, we see that higher abatement activities may raise or lower economic growth. The reason is as above. On the one hand, more abatement activities reduce investment spending; on the other hand, higher abatement activities have positive indirect growth effects by reducing the temperature increase, and thus the damage, and by reducing the value $k = K/H$.

To get further insights, we undertake simulations in the next section.

9.3 NUMERICAL EXAMPLES

We consider one time period to comprise one year. The population growth rate is assumed to be $n = 0.02$ and the depreciation rate of capital is $\delta = 0.075$. The preindustrial level of GHGs is normalized to one, i.e. $M_0 = 1$. We do this because M denotes all types of greenhouse gases (e.g. such as CO_2 and CH_4) so that we cannot insert the specific concentration of a certain type of greenhouse gas, say CO_2 or CH_4. Further, we are interested in the change of GHG concentration relative to the pre-industrial level. Further, we set $\gamma = 0.9$ which is motivated by an OECD study which runs regressions with emissions per capita as the dependent variable which is explained among others by GDP per capita and which obtains a value of about 0.9 (see OECD, 1995). β_1 and ξ are set to $\beta_1 = 1.1$ and $\xi = 0.3$ (see chapter 8). $c = 0.8$ and the tax share is set to $\tau = 0.2$ which is about equal to the tax share in Germany in 1996 (see Sachverständigenrat, 2001). The capital share is $\alpha = 0.35$ and A is set to $A = 2.9$. As to τ_b we consider the values $\tau_b = 0.0075, 0.01, 0.0125$. For example, in Germany the ratio of abatement spending to prevent air pollution to total tax revenue was 0.01 in 1996 (see Sachverständigenrat, 2001, table 30 and table 65). a is set to $a = 0.00075$. This implies that GHGs double for $\tau_b = 0.01$.

An important role is played by the damage functions $D(\cdot)$. This will be introduced now. As to $D(\cdot)$ we assume the function

$$D(\cdot) = \left(m_1 (T - T_o)^2 + 1 \right)^{-\phi}, \tag{9.13}$$

with $m_1 > 0, \phi > 0$.

As to the numerical values of the parameters in (9.13) we assume $m_1 = 0.05$ and $\phi = 0.05$ and $m_1 = 0.025$ and $\phi = 0.025$. ($m_1 = 0.05, \phi = 0.05$) implies that an increase of the surface temperature by 1 (2, 3) degree(s) leads to a decrease of aggregate production by 0.2 (0.9, 1.8) percent. The combination ($m_1 = 0.025, \phi = 0.025$) implies that an increase of the surface temperature by 1 (2, 3) degree(s) leads to a decrease of aggregate production by 0.06 (0.2, 0.5) percent. Comparing these values with the estimates published in IPCC (1996), we see that the values we choose yield a damage which is a bit lower than the one reported by IPCC (1996).

Table 9.1 Varying the Abatement Share
between 0.0075 and 0.0125 with
$m_1 = 0.05, \phi = 0.05$.

τ_b	T^*	M^*	g	g_Y
0.0075	293.0	2.63	0.0185	0.0191
0.01	291.8	2.03	0.0198	0.0203
0.0125	290.8	1.66	0.0207	0.0211

In the following tables we report the results of our numerical studies where we report the temperature on the BGP and the GHG concentration on the BGP. Further, we report the balanced growth rate denoted by g. g_Y is the average growth rate of per capita GDP on the transition path for the next 100 years, where initial conditions are set to $T(0) = 289$, $M(0) = 1.13$ and $k(0) = 8.1$. The growth rate of output is given by

$$\frac{\dot{Y}}{Y} = \alpha \frac{\dot{K}}{K} + (1 - \alpha)\frac{\dot{H}}{H} + \frac{D'(\cdot)}{D(\cdot)} \dot{T}.$$

In table 9.1 we vary the abatement share between $\tau_b = 0.0075$ and $\tau_b = 0.0125$.

This table shows that an increase in the ratio of abatement spending to tax revenue, and also to GDP, leads to both higher growth rates and to a smaller increase in GHG emissions and, as a consequence, to a smaller increase in average global surface temperature. This holds both for the long-run balanced growth rate as well as for the transition path for the next 100 years. In this case, the decline in investment caused by more abatement spending is compensated by the higher production resulting from a smaller damage since the temperature increase is smaller with higher abatement. The maximum growth rate is obtained then there is no increase in the average temperature implying that the damage is zero. This is achieved for τ_b about $\tau_b = 0.02$. This outcome, of course, depends on the specification of the damage function. This is shown in the next table, where we set $m_1 = 0.025$ and $\phi = 0.025$.

Table 9.2 shows that the growth rate first rises when abatement spending is increased but then declines when abatement is further increased, although the growth effects are very small. This is due to the smaller damage caused by the increase in average global surface temperature.

Next we consider the case of varying the tax share between 15 and 25 percent. The results for ($m_1 = 0.05, \phi = 0.05$) are shown in table 9.3.

Table 9.3 shows that raising the tax rate reduces GHG emissions and also the balanced growth rate and the GDP growth rate on the transition path. It is true that a higher tax revenue raises economic growth and abatement spending because the latter are always in fixed proportion to

Table 9.2 Varying the Abatement Share
between 0.0075 and 0.0125 with
$m_1 = 0.025, \phi = 0.025$.

τ_b	T^*	M^*	g	g_Y
0.0075	293.0	2.63	0.0218	0.0222
0.01	291.8	2.03	0.022	0.0223
0.0125	290.8	1.66	0.0219	0.0222

Table 9.3 Varying the Tax Share
between 0.15 and 0.25 with
$m_1 = 0.05, \phi = 0.05$.

τ	T^*	M^*	g	g_Y
0.15	293.0	2.63	0.0266	0.0274
0.2	291.8	2.03	0.0198	0.0203
0.25	290.8	1.66	0.0124	0.0127

the tax revenue. However, the negative direct growth effect of a higher tax share clearly dominates the positive growth effect of smaller damage due to less GHG emissions and a smaller increase in temperature.

We also studied the effects of varying abatement activities for different values of γ. To do so we set $\gamma = 0.5$ and $\gamma = 1.5$.[3] From a qualitative point of view, the results are the same as for $\gamma = 0.9$. The only difference is that the quantitative growth effects are a bit different.

In the next chapter we present a special form of the growth model by assuming that physical capital and the stock of knowledge can be summarized in one variable. This gives the so-called *AK* model of endogenous growth.

---

[3] To get temperature increase that are compatible with IPCC estimates, we set $a = 0.00035$ and $a = 0.0011$, respectively.

10

The *AK* Endogenous Growth Model

Assuming that physical capital and human capital evolve at the same rate, that is, $\dot{K} = \dot{H}$ and $K(0) = H(0)$ hold, allows us to rewrite the aggregate per capita production function as follows,

$$Y = AKD(\cdot), \tag{10.1}$$

which is linear in capital. In the economics literature, this simplifying assumption is frequently made, and in the following we study this model. First, we consider the descriptive growth model and then analyze the second-best solution where abatement activities are chosen optimally.

The differential equation describing the evolution of capital is now given by

$$\dot{K} = AKD(\cdot)(1 - \tau(1 + \tau_b) - c(1 - \tau)) - (\delta + n)K, \tag{10.2}$$

where we assume again $B = \tau_b X$ and $C = cY(1 - \tau)$ as in section 9.1. The economy then is completely described by (10.2), (8.5), and (8.6), with emissions given by (9.9), and the balanced growth path (BGP) is

$$g = AD(\cdot)(1 - \tau(1 + \tau_b) - c(1 - \tau)) - (\delta + n). \tag{10.3}$$

The balanced growth rate is independent from the capital stock but only depends on the average global surface temperature T. This implies that a BGP is as defined in section 9.2 with the only difference that we only have to consider the equations $\dot{M}$ and $\dot{T}$.[1]

From (10.3) it is immediately seen that variations in τ_b and in τ affect the balanced growth rate both directly as well as indirectly by affecting the temperature on the balanced growth path. As in chapter 9, there is a positive indirect growth effect and a negative direct growth effect going along with changes in τ_b and in τ; the overall effect, however, cannot be determined for the general model. Therefore, we present the results of numerical examples without presenting the results for the analytical model.

The parameters are as in section 9.3, that is, $n = 0.02, \delta = 0.075, \tau = 0.2,$ $M_0 = 1, c = 0.8, \gamma = 0.9, \beta_1 = 1.1, \xi = 0.3, a = 0.00075,$ and $\alpha = 0.35$. The

---

[1] Of course, the *AK* model is also asymptotically stable.

Table 10.1 Varying the Abatement Share between 0.0075 and 0.02 with $\tau = 0.2$.

τ_b	T^*	M^*	g
0.0075	293.0	2.63	0.0197
0.01	291.8	2.03	0.0208
0.0125	290.8	1.66	0.0216
0.018	289.3	1.19	0.022
0.02	288.8	1.08	0.0219

Table 10.2 Varying the Tax Share 0.15 and 0.25 with $\tau_b = 0.01$.

τ	T^*	M^*	g
0.15	293.0	2.63	0.0269
0.2	291.8	2.03	0.0208
0.25	290.8	1.66	0.0142

only different parameter is the value of A, which we set to $A = 0.75$ to get growth rates that are in line with those observable in the real world. The tax share is set to $\tau = 0.2$, and we consider for τ_b the values $\tau_b = 0.0075, 0.01, 0.0125$. m_1 and ϕ are again set to $m_1 = 0.05$ and $\phi = 0.05$.

Tables 10.1 and 10.2 present the results of varying abatement spending and of variations of the tax share with $\tau = 0.2$, and $\tau_b = 0.01$, respectively.[2]

Tables 10.1 and 10.2 largely confirm the results of chapter 9. That is, a rise in the tax share reduces the balanced growth rate and also GHG emissions. The growth rates in the AK model differ a bit from those of chapter 9, but the change is about the same. However, in contrast to chapter 9, the growth rate is maximized for τ_b such that the temperature is larger than the preindustrial temperature T_0. Next, we consider the second-best solution.

10.1 THE SECOND-BEST SOLUTION

To derive the second-best solution we assume that the government takes private consumption and the tax share as given and sets abatement such that welfare is maximized. As to welfare, we assume as usual

---

[2] In this section we only consider the balanced growth path.

that it is given by the discounted stream of per capita utility times the number of individuals over an infinite time horizon. Concerning utility, we assume a logarithmic function. More concretely, the government solves the following optimization problem

$$\max_{\tau_b} \int_0^\infty e^{-(\rho-n)t} L(0) \ln(c(1-\tau)AKD(\cdot)) dt \qquad (10.4)$$

subject to (10.2), (8.6), (8.5) with $c(1-\tau)AKD(\cdot) = C$ per capita consumption. ln denotes the natural logarithm, and ρ is the discount rate. In the following we normalize $L(0) \equiv 1$.

To find necessary optimality conditions, we formulate the current-value Hamiltonian as

$$\mathcal{H}(\cdot) = \ln(c(1-\tau)AKD(\cdot)) + \lambda_1(AKD(\cdot)(1-\tau(1+\tau_b)) - c(1-\tau))$$

$$- (\delta+n)K) + \lambda_2 \left(\beta_2 \left(\frac{a}{\tau_b \tau} \right)^\gamma - \mu M \right) + \lambda_3 (c_h)^{-1} \cdot \left(\frac{1367.5}{4} 0.21 \right.$$

$$\left. -(5.67 \, 10^{-8})(19.95/109)T^4 + \beta_1 (1-\xi) 6.3 \ln \frac{M}{M_0} \right), \qquad (10.5)$$

with λ_i, $i = 1,2,3$, the shadow prices of K, M, and T, respectively, and $E = a^\gamma Y^\gamma B^{-\gamma}$ emissions. Note that λ_1 is positive while λ_2 and λ_3 are negative.

The necessary optimality conditions are obtained as

$$\frac{\partial \mathcal{H}(\cdot)}{\partial \tau_b} = \lambda_1 AKD(\cdot)(-\tau) - \lambda_2 \beta_2 (a/\tau)^\gamma (-\gamma) \tau_b^{-\gamma-1} = 0, \qquad (10.6)$$

$$\dot{\lambda}_1 = (\rho+\delta)\lambda_1 - K^{-1} - \lambda_1 AD(\cdot)(1-\tau(1+\tau_b)) - c(1-\tau)), \quad (10.7)$$

$$\dot{\lambda}_2 = (\rho-n)\lambda_2 + \lambda_2 \mu - \lambda_3 (1-\xi) \beta_1 6.3 c_h^{-1} M^{-1}, \qquad (10.8)$$

$$\dot{\lambda}_3 = (\rho-n)\lambda_3 - \frac{D'(\cdot)}{D} - \lambda_1 AKD'(\cdot)(1-\tau(1+\tau_b)) - c(1-\tau))$$

$$+ \frac{\lambda_3 (5.67 \, 10^{-8}(19.95/109) 4 T^3)}{c_h}. \qquad (10.9)$$

Furthermore, the limiting transversality condition $\lim_{t\to\infty} e^{-(\rho+n)t}(\lambda_1 K + \lambda_2 T + \lambda_3 M) = 0$ must hold.

From (10.6) we get the second-best optimal abatement activities (as ratio to the tax revenue) as

$$\tau_b^o = \left(\frac{-\lambda_2 \beta_2 \gamma (a/\tau)^\gamma}{AD(\cdot)K\lambda_1 \tau} \right)^{1/(1+\gamma)}. \qquad (10.10)$$

Equation (10.10) shows that τ_b^o is higher the more polluting the technology in use is, which is modeled in the framework by the coefficient a.

This means that economies with less clean production technologies have a higher optimal abatement share than economies with a cleaner technology. However, this does not mean that economies with a cleaner technology have higher emissions. This holds because on one hand, the higher abatement share may not be high enough to compensate for the more polluting technology. On the other hand, the second-best pollution tax rate also depends on λ_1, λ_2, and K. Further, from the expression for τ_b^o one realizes that the higher the absolute value of the shadow price of GHG concentrations, $|\lambda_2|$, the higher the abatement share has to be set.

For the second-best solution, a balanced growth path is defined similar to definition 4.

Definition 5 *For the second-best solution, a balanced growth path is a path such that* $\dot{T} = \dot{M} = \dot{\lambda}_2 = \dot{\lambda}_3 = 0$ *and* $\dot{K}/K = -\dot{\lambda}_1/\lambda_1$ *hold, with* $M \geq M_o$.

Unlike for the descriptive versions of our growth models, we cannot give results as to the existence and stability of a BGP for the analytical model. Therefore, and to get an impression about the quantitative results of the second-best solution, we again make simulations with the same parameters as in chapter 8, with $\tau = 0.2$ and a discount rate of 5 percent, $\rho = 0.05$. For the numerical values of the parameters, it can be shown that there exists a unique BGP that is, however, not asymptotically stable but a saddle point. This is the content of proposition 17.

Proposition 17 *For the second-best model economy there exists a unique BGP that is a saddle point for the numerical parameter values of section 9.3.*

Proof: To prove that proposition we define $\Lambda \equiv K \cdot \lambda_1$ giving $\dot{\Lambda}/\Lambda = \dot{K}/K + \dot{\lambda}_1/\lambda_1$. Setting $\dot{\Lambda}/\Lambda = 0$ gives $\Lambda^* = (\rho - n)^{-1}$. Inserting Λ^* in (10.10) and the resulting expression as well as Λ^* in equations (8.6), (8.5), (10.8), and (10.9) gives an autonomous system of differential equations that depends on M, T, λ_2, and λ_3. A rest point of this system yields a BGP. With the numerical parameters the solution is given by $M^* = 1.25625$, $T^* = 289.50603$, $\lambda_2^* = -0.75023$, and $\lambda_3^* = -0.00378$.

The eigenvalues of the Jacobian are given by

$$e_1 = 6.75544, \quad e_2 = -6.75544 \quad e_3 = 0.19010 \quad e_4 = -0.19010.$$

Thus, the proposition is proved. For $a = 5 \cdot 10^{-4}$ the numerical values are different, but the qualitative result remains unchanged. $\square$

Table 10.3 gives the balanced growth rate and the values of T, M, B/Y, and τ_b^o on the BGP.

Table 10.3 shows that for the AK growth model there exists an optimal share for abatement spending that maximizes utility and also the balanced growth rate. This share is about 1.7 percent of the tax revenue for $a = 7.5 \cdot 10^{-4}$, implying a share of abatement spending per GDP, B^*/Y^*,

Table 10.3 The Second-Best Solution.

a	τ_b^o	B^*/Y^*	T^*	M^*	g
$7.5 \cdot 10^{-4}$	0.017	0.0034	289.5	1.26	0.0221
$5 \cdot 10^{-4}$	0.012	0.0024	289.2	1.17	0.0229

of 0.34 percent. Further, it is seen that in a world with a more polluting technology (higher a) the second-best pollution tax rate is larger, but nevertheless, emissions and the increase in temperature are also higher. Economies with a more polluting production technology should have have a higher pollution tax rate. In spite of this, the emissions in the economy with the more polluting technology are higher because the higher abatement share is not enough to compensate for the more polluting technology.

Thus, our model is in part consistent with the literature that postulates that an environmental Kuznets curve exists, where emissions first rise with an increase in GDP (when the technology in use is relatively polluting) but decline again when a certain level of GDP is reached and the technology becomes cleaner (see, e.g., the contribution by Stockey 1998).[3] But keep in mind that our result is obtained for second-best government policies and it may be doubted that in reality governments pursue optimal policies.

It should also be mentioned that the choice of the discount rate ρ affects the optimal τ_b^o and also the balanced growth rate. However, neither the growth rate nor optimal abatement react sensitively to changes in ρ. For example, setting $\rho = 0.02$ or $\rho = 0.1$ basically leaves unchanged both optimal abatement spending and the growth rate.

In the next section we study the problem of the social planner. The difference to the optimization problem faced by the government in this section is that the social planner decides on both the consumption share and the abatement share instead of taking consumption as given.

10.2 THE SOCIAL OPTIMUM

As mentioned at the end of the last section, the social planner can decide on both the investment share and the abatement share. The optimization problem, then, is given by

$$\max_{c_s,b} \int_0^\infty e^{-(\rho-n)t} L(0) \ln(c_s \, AKD(\cdot)) dt, \qquad (10.11)$$

---

[3] In the model by Stockey, the pollution intensity is a choice variable, whereas it is a parameter in our model.

with c_s the consumption share and b the abatement share in the social optimum, respectively. Again, ln denotes the natural logarithm, ρ is the discount rate, and we normalize $L(0) \equiv 1$. The constraints are (8.5) and (8.6), where $b \equiv B/Y$, and the differential equation $\dot{K}$. The latter is now given by $\dot{K} = A K D(\cdot)(1 - c_s - b) - (\delta + n)K$. Again, we formulate the current-value Hamiltonian, which is

$$\mathcal{H}(\cdot) = \ln(c_s \, AKD(\cdot)) + \lambda_4 \dot{K} + \lambda_5 \dot{M} + \lambda_6 \dot{T}, \qquad (10.12)$$

with λ_i, $i = 4, 5, 6$, the shadow prices of K, M, and T in the social optimum. As in the previous section, λ_4 is positive and λ_5 and λ_6 are negative.

The necessary optimality conditions now are

$$\frac{\partial \mathcal{H}(\cdot)}{\partial c_s} = c_s^{-1} - \lambda_4 AKD(\cdot) = 0, \qquad (10.13)$$

$$\frac{\partial \mathcal{H}(\cdot)}{\partial b} = -\lambda_4 AKD(\cdot) - \lambda_5 \beta_2 \gamma b^{-\gamma-1} a^\gamma = 0, \qquad (10.14)$$

$$\dot{\lambda}_4 = (\rho + \delta)\lambda_4 - K^{-1} - \lambda_4 \, A D(\cdot)(1 - c_s - b), \qquad (10.15)$$

$$\dot{\lambda}_5 = (\rho - n + \mu)\lambda_5 - \lambda_6 (1 - \xi)\,\beta_1\, 6.3\, c_h^{-1} M^{-1}, \qquad (10.16)$$

$$\dot{\lambda}_6 = (\rho - n)\lambda_6 - \frac{D'(\cdot)}{D} - \lambda_4 \, A K D'(\cdot)(1 - c_s - b)$$

$$+ \frac{\lambda_6 \left(5.67 \, 10^{-8} (19.95/109) \, 4\, T^3\right)}{c_h}. \qquad (10.17)$$

The transversality condition is as in the last section.

From (10.13) and (10.14) we get the first-best consumption share and abatement share (relative to GDP, respectively) as

$$c_s = (\lambda_4 AKD(\cdot))^{-1} \qquad (10.18)$$

$$b = \left(\frac{-\lambda_5 \beta_2 \gamma a^\gamma}{AD(\cdot)K\lambda_4}\right)^{1/(1+\gamma)}. \qquad (10.19)$$

One immediately realizes that (10.19) is similar to (10.10) and the interpretation is basically the same as in 10.1. Equation (10.18) shows that the first-best optimal consumption share negatively depends on both physical capital and its shadow price. That means the higher the stock of capital and the higher its "price," the smaller the consumption share. Of course, as for the second-best solution, consumption grows while the consumption share is constant on the BGP. Furthermore, the higher the damage caused by the temperature increase, the smaller the consumption share in the economy. This outcome is due to the fact that the temperature increase negatively affects

Table 10.4 The Social Optimum.

a	B^*/Y^*	T^*	M^*
$7.5 \cdot 10^{-4}$	0.0041	288.65	1.05
$5 \cdot 10^{-4}$	0.0028	288.57	1.04

aggregate production, Y, and thus consumption, which is equal to $c_s \cdot Y$.

As to the question of whether there exists a BGP for the social optimum, we again resort to simulations. Here, we can state the following proposition.

Proposition 18 *For the social optimum there exists a unique BGP, defined analogous to definition 5, which is a saddle point for the numerical parameter values in section 9.3.*

Proof: The proof of the proposition proceeds in complete analogy to that of proposition 17. Therefore, we do not mention it in detail. Again, the value of a affects the numerical values but leaves the qualitative outcome unchanged. □

In the following we focus our attention on the abatement share and the temperature increase in the social optimum compared to the second-best economy. Table 10.4 gives the values of T, M, and B/Y on the BGP for $a = 7.5 \cdot 10^{-4}$ and $a = 5 \cdot 10^{-4}$, respectively. The other parameters are as in section 10.1.

Table 10.4 demonstrates that, in the social optimum, economies with a less polluting technology (smaller a) have a smaller abatement share than economies with a more polluting technology. Nevertheless, the economies with the cleaner technology have a smaller level of GHG emissions and, consequently, a smaller increase in temperature. This result is equivalent to the one obtained for the second-best solution. That means that the higher abatement share cannot compensate for the less clean production technology. One also realizes that the abatement share in the social optimum is higher compared to the second-best solution, and as a consequence, the temperature increase is smaller.

10.3 A MULTIREGION WORLD

Up to now we have assumed that the world consists of homogeneous regions that can be represented by one economy. In this section we allow for regions that differ with respect to the damages they suffer from global warming and with respect to their emissions (see Greiner 2005b).

We assume that aggregate production in region i, $i = 1, \ldots, n$, takes place according to the following per capita production function:

$$Y_i(t) = A_i K_i D_i(T - T_o), \qquad (10.20)$$

with Y_i per capita production in region i, A_i a positive constant, and K_i a composite of human and physical capital. $D_i(T - T_o)$ is the function giving the decline in aggregate per capita production in country i resulting from deviations of the actual temperature from the preindustrial temperature, T_o.

We point out that AK models are very sensitive with respect to the parameters. However, we do not intend to make calibrations, but we want to get insights into the structure of the model and see how certain climate policies affect economies qualitatively. This should be kept in mind when interpreting the results in the next sections.

As to the function $D_i(T - T_o)$, we assume that it is continuously differentiable and that it satisfies (9.5) and (9.6).

The per capita capital accumulation function is given by

$$\dot{K}_i = A_i K_i D_i(\cdot)(1 - c_i - \tau_{B,i}) - (\delta_i + n_i)K_i, \qquad (10.21)$$

with c_i the consumption share in region i and $\tau_{B,i}$ the abatement share. $n_i \in (0, 1)$ is the population growth rate in region i and $\delta_i \in (0, 1)$ is the depreciation rate of capital.

Recall that we take as a starting point the Solow-Swan approach with a given consumption and saving share. We do this because we want to focus on effects resulting from climate changes that affect production as modeled in equations (10.20)–(10.21) and therefore neglect effects resulting from different preferences.

From equations (10.20) and (10.21) we see that the gross marginal product of private capital, which equals the interest rate in our economy, is equal to $A_i D_i(\cdot)$ and that deviations from the preindustrial temperature affect both the level of production as well as the growth rate of capital and production.

Concerning GHG emissions, we again assume that these are a byproduct of production and expressed in CO_2 equivalents. Emissions are a function of per capita output relative to per capita abatement activities. This implies that higher production goes along with more emissions for a given level of abatement spending. It should also be mentioned that the emission of GHGs does not affect production directly but only indirectly by raising the concentration of GHGs in the atmosphere, which affects the climate of the Earth and leads to a higher surface temperature and more extreme weather situations.

As far as emissions go, we make the same assumption as in section 9.1. The only difference now is that we consider different regions. Formally, emissions in region i are described by

$$E_i = \left(\frac{a_i Y_i}{B_i}\right)^{\gamma_i} = \left(\frac{a_i}{\tau_{B,i}}\right)^{\gamma_i}, \quad (10.22)$$

where B_i is per capita abatement with $B_i = \tau_{B,i} Y_i$. $\gamma_i > 0$ and $a_i > 0$ are positive constants. The parameter a_i can again be interpreted as a technology index describing how polluting a given technology is. For large values of a_i, a given production (and abatement) goes along with high emissions, implying a relatively polluting technology and vice versa.

The concentration of GHGs, M, then evolves according to the following differential equation

$$\dot{M} = \beta_1 \sum_{j=1}^{n} E_j - \mu M, M(0) = M_0. \quad (10.23)$$

Note that this is analogous to section 9.1, and we also assume the same parameter values as in 9.1. The equation giving the change in global average surface temperature is given by equation (8.5).

Thus, the world economy is completely described by equations (10.21), (8.5), and (10.23), with emissions given by (10.22).

10.3.1 The Noncooperative World

In this section we analyze the noncooperative world or the Nash equilibrium. Each region maximizes utility resulting from per capita consumption where we assume a logarithmic utility function. Assuming that each region maximizes per capita utility, the optimization problem in each region $i = 1, \ldots, n$ is given by

$$\max_{\tau_{B,i}} \int_0^\infty e^{-\rho_i t} \ln(c_i A_i K_i D_i(\cdot)) dt, \quad (10.24)$$

subject to (10.23), (8.5), and (10.21) with $c_i A_i K_i D_i(\cdot) = C_i$ per capita consumption. ln denotes the natural logarithm, and ρ_i is the discount rate.

To find the optimum we construct the current-value Hamiltonian, which is

$$\mathcal{H}_i(\cdot) = \ln(c_i A_i K_i D_i(\cdot)) + \lambda_{1,i} \left(\beta_1 \sum_{j=1}^{n} \left(\frac{a_j}{\tau_{B,j}}\right)^{\gamma_j} - \mu M \right)$$

$$+ \lambda_{2,i} \left(k_1 - k_2 T^4 + k_3 \ln \frac{M}{M_0} \right)$$

$$+ \lambda_{3,i} (A_i K_i D_i(\cdot)(1 - c_i - \tau_{B,i}) - (\delta_i + n_i) K_i), \quad (10.25)$$

with $k_1 \equiv c_h^{-1} 0.21 \cdot 1367.5/4$, $k_2 \equiv c_h^{-1} 0.95 \left(5.67\ 10^{-8}\right) (21/109)$, and $k_3 \equiv$ $4.851 c_h^{-1}$. $\lambda_{k,i}$, $i = 1,2,3$, denote the shadow prices of M, T, and K_i in region i, respectively, and $E_i = a_i^{\gamma_i} Y_i^{\gamma_i} B_i^{-\gamma_i}$ are emissions. Note that $\lambda_{1,i}$ and $\lambda_{2,i}$ are negative, and $\lambda_{3,i}$ are positive.

The necessary optimality conditions are obtained as

$$\frac{\partial \mathcal{H}_i(\cdot)}{\partial \tau_{B,i}} = \lambda_{1,i} \beta_1 (-\gamma_i) a_i^{\gamma_i} \tau_{B,i}^{-\gamma_i - 1} - \lambda_{3,i} A_i K_i D_i(\cdot) = 0, \tag{10.26}$$

$$\dot{\lambda}_{1,i} = (\rho_i + \mu)\, \lambda_{1,i} - \lambda_{2,i} k_3 M^{-1}, \tag{10.27}$$

$$\dot{\lambda}_{2,i} = \rho_i \lambda_{2,i} - D_i'(\cdot)/D_i + \lambda_{2,i} k_2\, 4\, T^3$$
$$\qquad - \lambda_{3,i} A_i K_i D_i'(\cdot)(1 - c_i - \tau_{B,i}), \tag{10.28}$$

$$\dot{\lambda}_{3,i} = (\rho_i + \delta_i + n_i)\, \lambda_{3,i} - K_i^{-1} - \lambda_{3,i} A_i D_i(\cdot)(1 - c_i - \tau_{B,i}). \tag{10.29}$$

Furthermore, the limiting transversality condition $\lim_{t \to \infty} e^{-\rho_i t}(\lambda_{1,i} M + \lambda_{2,i} T + \lambda_{3,i} K_i) = 0$ must hold.

From (10.26) we get the optimal abatement activities (as a ratio to GDP) in each region as

$$\tau_{B,i}^o = \left(\frac{\beta_1 (-\lambda_{1,i}) \gamma_i a_i^{\gamma_i}}{\lambda_{3,i} A_i K_i D_i(\cdot)} \right)^{1/(1+\gamma_i)}. \tag{10.30}$$

Equation (10.30) shows that $\tau_{B,i}^o$ is higher the more polluting the technology in use is, which is modeled in our framework by the coefficient a_i. This means that economies with less clean production technologies have a higher optimal abatement share than these with a cleaner technology. However, this does not mean that economies with a cleaner technology have higher emissions, which was already obtained in the last section. This holds because on one hand, the higher abatement share may not be high enough to compensate for the more polluting technology. On the other hand, the second-best pollution tax rate also depends on $\lambda_{1,i}$, $\lambda_{3,i}$, and K_i. Further, from the expression for $\tau_{B,i}^o$ one realizes that the higher the absolute value of the shadow price of the GHG concentration, $|\lambda_{1,i}|$, the higher the abatement share has to be set.

In the following we confine our investigations to the BGP. A BGP is defined analogously to the definition 4.[4] In particular, we require again that the GHG concentration and the temperature must be constant along a BGP (sustainability aspect). Furthermore, the growth rate of per capita capital is constant over time. Finally, for reasons of realism, we only consider BGPs with an aggregate GHG concentration larger than or equal to the preindustrial level.

---

[4] In the following, *steady state* is again used equivalently to *balanced growth path*.

To gain further insight into our model, we use numerical calculations and consider three regions. Two are relatively highly developed
regions, where one is producing with a relatively clean technology and
the other uses a relatively polluting technology. One may think of the
European OECD countries as the first region and the United States as
the second region. The third region is given by low-income countries
with a technology that is more polluting than the other two regions. We
set $a_1 = 3.75\,10^{-4}$. a_2 is twice as large as a_1, that is, $a_1 = 7.5\,10^{-4}$, and
a_3 is four times as large as a_1, that is, $a_3 = 0.003$. These relations reflect
about the situation in European OECD countries relative to the United
States and relative to low-income countries in 1995 (see Nordhaus and
Boyer 2000, table 3.1). γ_i, $i = 1, 2, 3$, is set to one in all three regions, that
is, $\gamma_i = 1, i = 1, 2, 3$.

As to the function $D_i(T(t) - T_o)$ we assume the specification as in 9.1,

$$D_i = \left(1 + m_i(T - T_o)^2\right)^{-\phi_i}, \quad m_i, \phi_i > 0. \tag{10.31}$$

The damage caused by a higher GHG concentration is assumed to be the
same for the first and second region and about three times as high in the
third region for a doubling of GHGs. This is achieved by the following
parameter values, $m_1 = m_2 = 0.0013$, $\phi_1 = \phi_2 = 1$, and $m_3 = 0.0087$,
$\phi_3 = 0.5$. This implies that an increase of the average surface temperature
by 3 °C as a result of a doubling of GHGs goes along with a damage of
about 1.2 percent in regions 1 and 2. A rise of the temperature by about
6 °C implies a damage of roughly 4.5 percent. For the third region the
damage is 4 percent for a 3 °C increase of the temperature and about
13 percent when the temperature rises by 6 °C. These values roughly
reflect the situation in European OECD countries, the United States,
and low-income countries (see Hackl and Pruckner 2003, table 1).

Damages are not the same in the regions because of differences in the
state of development. For example, in developing countries people are
prepared worse for possible catastrophes than in developed countries
because they cannot afford to invest in preventive measures. Furthermore, poor countries depend more heavily on agriculture and have
less means to compensate losses in agricultural production compared to
more developed countries so that the consequences of climatic changes
are more dramatic in less developed countries.

The subjective discount rate is assumed to be the same in the three
regions and we set $\rho_i = 0.03$, $i = 1, 2, 3$. We assume the same discount
rates in all regions because we want to focus on growth effects resulting from the supply side, which is affected by a possible temperature
increase, and we are not interested in differences resulting from different
preferences. If the discount rates were different, this would lead to differences in growth rates even if the effects of the temperature increase in
the regions were the same, and this would complicate the analysis. The

population growth rates are assumed to be zero in the first two regions, $n_1 = n_2 = 0$, and 2 percent in the third region, $n_3 = 0.02$.

The marginal propensity to consume is set to 80 percent in all three regions, $c_i = 0.8, i = 1, 2, 3$. The marginal product of capital in the second region is assumed to be larger than in the first region, and the latter is larger than in the third region, and we set $A_1 = 0.35$, $A_2 = 0.5$, and $A_3 = 0.25$. This implies a higher marginal product of capital in the second region compared with the first and third. A justification for different marginal products can be seen in different levels of technology and in capital transfer constraints. Depreciation rates are set to $\delta_1 = \delta_2 = 0.04$ in regions 1 and 2 and $\delta_3 = 0.01$ in region 3. Thus, we acknowledge that depreciation of capital is higher in those regions with higher income.

Defining $\kappa_i \equiv \lambda_{3,i} \cdot K_i$, a BGP is given by the solution of the equations.

$$0 = \beta_1 \sum_{j=1}^{3} \left(\frac{a_j}{\tau_{B,j}^o} \right) - \mu M, \tag{10.32}$$

$$0 = k_1 - k_2 T^4 + k_3 \ln \frac{M}{M_o}, \tag{10.33}$$

$$0 = \kappa_i \left(\dot{K}_i / K_i + \dot{\lambda}_{3,i} / \lambda_{3,i} \right), \tag{10.34}$$

$$0 = (\rho_i + \mu) \lambda_{1,i} - \lambda_{2,i} k_3 M^{-1}, \tag{10.35}$$

$$0 = \rho_i \lambda_{2,i} - D_i'(\cdot)/D_i + \lambda_{2,i} k_2 4 T^3 - \kappa_i A_i D_i'(\cdot)(1 - c_i - \tau_{B,i}), \tag{10.36}$$

with $\tau_{B,i}^o = \left((\beta_1(-\lambda_{1,i})a_i)/(\lambda_{3,i} A_i K_i D_i(\cdot)) \right)^{0.5}$, $i = 1, 2, 3$. Equation (10.32) follows from (10.23), and (10.33) follows from (8.5). Equation (10.34) is obtained by combining (10.21), and (10.29), and (10.35) and (10.36), finally, are obtained from (10.27) and (10.28). It should be noted that a constantly rising capital stock goes along with a constantly declining (shadow) price of capital, implying that κ_i is constant on a BGP. Solving equations (10.32)–(10.36) gives steady-state values[5] for the level of GHGs ($M^\star$), the temperature ($T^\star$), the product of the capital stock and its shadow price ($\kappa^\star$), and the shadow prices of GHGs ($\lambda_1^\star$) and temperature ($\lambda_2^\star$). These variables, then, give the balanced growth rate in region i, which is given by $g_i \equiv A_i D_i(\cdot)(1 - c_i - \tau_{B,i}) - (\delta_i + n_i)$, with $\tau_{B,i}^o$ as before.

As for the existence of a BGP, Rosen (1965) has derived general conditions such that an N-person game has a unique solution. However, this result cannot be applied to our model because we consider a differential game with ongoing growth. Therefore, we will numerically compute BGPs in our examples where the computations show that existence and local uniqueness is assured in each case.

---

[5] Recall that the $\star$ denotes values on the BGP.

Table 10.5 Optimal Abatement Shares, Emissions, Balanced Growth
Rates for the three Regions as Well as Average Global
Temperature (Noncooperative Case) for $\mu = 0.1$. (The values
in parantheses are for $\mu = 0.005$.)

$\tau_{B,1}^o$	E_1	g_1	$\tau_{B,2}^o$	E_2	g_2	$\tau_{B,3}^o$	E_3	g_3
0.29%	0.131	2.69%	0.38%	0.1949	5.51%	1.36%	0.2214	1.27%
(0.15%)	(0.25)	(0.5%)	(0.2%)	(0.377)	(2.4%)	(0.7%)	(0.443)	(−0.8%)

$T^* = 293.1 \ (308.8)$

In table 10.5 we give the result of our calculations for the three regions. As to the rate of decay of GHGs, we consider two values, $\mu = 0.1$ and $\mu = 0.005$.

For $\mu = 0.1$ this table shows that the region with the less clean production technology (region 2) has a higher abatement share than the region with the cleaner production technology (region 1) if damages caused by a rise in the average surface temperature are the same in both regions. However, this does not mean that emissions in region 2 are smaller than in region 1. So region 1 has fewer emissions than region 2. This means that the higher abatement share cannot compensate for the less clean production technology.

Taking into account that both the production technology and the damages caused by a rise in GHGs are different (comparing regions 2 and 3), one can see that region 2 spends relatively less for abatement than region 3–0.4 percent versus 1.4 percent. Further, region 3 has higher emissions than region 2, although the first spends a higher share of GDP for abatement.

With no cooperation, GHGs rise by about 2.7 of the preindustrial level, implying an increase in the average global surface temperature of 4.7 °C for a decay rate of GHGs of 10 percent and the other parameter values we assume.

Setting $\mu = 0.005$, the qualitative results remain unchanged. It should also be mentioned that now the model does not produce sustained growth in region 3 because of the high damages going along with the rise in temperature. The temperature increase in this case is about 10 °C.

Note that we study optimal policies. Therefore, the outcome of the numerical example does not necessarily reflect what we can observe in reality. This holds because abatement policies in real-world economies are not necessarily optimal.

In the next section we compare this result to the outcome in the cooperative world.

10.3.2 The Cooperative World

In the cooperative world, the optimization problem of the planner is given by[6]

$$\max_{\tau_{B,i}} \int_0^\infty e^{-\rho t} \sum_{j=1}^n w_j \ln(c_j A_j K_j D_j(\cdot)) dt, \tag{10.37}$$

subject to (10.23) and (10.21) with $c_i A_i K_i D_i(\cdot) = C_i$ per capita consumption in region i. ln again denotes the natural logarithm, and ρ is the discount rate. w_i gives the weight given to region i.

To find the optimum, we construct the current-value Hamiltonian, which is now written as

$$\mathcal{H}(\cdot) = \sum_{j=1}^n w_j \ln(c_j A_j K_j D_j(\cdot)) + \lambda_4 \left(\beta_1 \sum_{j=1}^n \left(\frac{a_j}{\tau_{B,j}} \right)^{\gamma_j} - \mu M \right)$$

$$+ \lambda_5 \left(k_1 - k_2 T^4 + k_3 \ln \frac{M}{M_0} \right)$$

$$+ \sum_{j=1}^n \lambda_{6,j} (A_j K_j D_j(\cdot)(1 - c_j - \tau_{B,j}) - (\delta_j + n_j)K_j), \tag{10.38}$$

with λ_4, λ_5 the shadow prices of M and T and $\lambda_{6,i}$ the shadow prices of K_i. Again, λ_4 and λ_5 are negative and $\lambda_{6,i}$ positive.

The necessary optimality conditions are obtained as

$$\frac{\partial \mathcal{H}(\cdot)}{\partial \tau_{B,i}} = \lambda_4 \beta_1 (-\gamma_i) a_i^{\gamma_i} \tau_{B,i}^{-\gamma_i - 1} - \lambda_{6,i} A_i K_i D_i(\cdot) = 0, \tag{10.39}$$

$$\dot{\lambda}_4 = (\rho + \mu) \lambda_4 - \lambda_5 k_3 M^{-1}, \tag{10.40}$$

$$\dot{\lambda}_5 = \lambda_5 \rho + \lambda_5 k_2 4 T^3 - \sum_{j=1}^n w_j D_j'(\cdot)/D_j$$

$$- \sum_{j=1}^n \lambda_{6,j} A_j K_j D_j'(\cdot)(1 - c_j - \tau_{B,j}), \tag{10.41}$$

$$\dot{\lambda}_{6,i} = (\rho + \delta_i + n_i) \lambda_{6,i} - w_i K_i^{-1} - \lambda_{6,i} A_i D_i(\cdot)(1 - c_i - \tau_{B,i}). \tag{10.42}$$

Further, the limiting transversality condition $\lim_{t\to\infty} e^{-\rho t}(\lambda_4 M + \lambda_5 T + \sum_{j=1}^n \lambda_{6,j} K_j) = 0$ must hold.

---

[6] We do not call this situation pareto optimum because in the pareto optimum the social planner would also determine the savings rate, which is exogenous in our context. Therefore, this solution is in a way second-best.

From (10.39) we get the optimal abatement ratios as

$$\tau_{B,i}^o = \left(\frac{\beta_1(-\lambda_4)\gamma_i a_i^{\gamma_i}}{\lambda_{6,i} A_i K_i D_i(\cdot)} \right)^{1/(1+\gamma_i)}. \tag{10.43}$$

Equation (10.43) basically is equivalent to (10.30) with the exception that the shadow prices are different. This holds because in the cooperative world regions do not optimize separately.

To get further insight we proceed as in the last section. That is, we consider three regions, insert numerical values for the parameters, and then calculate the corresponding abatement shares, emissions, balanced growth rates, as well as the rise in GHGs and in the average global surface temperature. The parameter values are as in the last section, with $\rho = 0.03$.

Defining $\kappa_i \equiv \lambda_{6,i} \cdot K_i$ a BGP is given by the solution of the following system of equations:

$$0 = \beta_1 \sum_{j=1}^{3} \left(\frac{a_j}{\tau_{B,j}^o} \right) - \mu M, \tag{10.44}$$

$$0 = k_1 - k_2 T^4 + k_3 \ln \frac{M}{M_o}, \tag{10.45}$$

$$0 = \kappa_i \left(\dot{K}_i/K_i + \dot{\lambda}_{6,i}/\lambda_{6,i} \right), \tag{10.46}$$

$$0 = (\rho + \mu)\lambda_4 - \lambda_5 k_3 M^{-1}, \tag{10.47}$$

$$0 = \lambda_5\rho + \lambda_5 k_2 \, 4 \, T^3 - \sum_{j=1}^{n} w_j \, D_j'(\cdot)/D_j - \sum_{j=1}^{n} \kappa_j \, A_j \, D_j'(\cdot)(1 - c_j - \tau_{B,j}), \tag{10.48}$$

with $\tau_{B,j}^o$ given by (10.43). Table 10.6 gives the result assuming equal weight to each region ($w_1 = w_2 = w_3 = 1$).

Table 10.6 Optimal Abatement Shares, Emissions, and Balanced Growth Rates for the Three Regions as Well as Average Global Temperature (Cooperative Case) for $\mu = 0.1$. (The values in parentheses are for $\mu = 0.005$.)

$\tau_{B,1}^o$	E_1	g_1	$\tau_{B,2}^o$	E_2	g_2	$\tau_{B,3}^o$	E_3	g_3
0.57%	0.065	2.75%	0.68%	0.11	5.59%	1.9%	0.155	1.41%
(0.39%)	(0.096)	(0.9%)	(0.5%)	(0.162)	(3.1%)	(1.5%)	(0.196)	(−0.5%)

$T^* = 290.7 \ (305)$

Table 10.7 Abatement Shares, Emissions, and Balanced Growth Rates for the Three Regions as Well as Average Global Temperature ($w_3 = 2w_1 = 2w_2 = 2, \mu = 0.1$).

$\tau_{B,1}^o$	E_1	g_1	$\tau_{B,2}^o$	E_2	g_2	$\tau_{B,3}^o$	E_3	g_3	T^*
0.66%	0.057	2.71%	0.78%	0.096	5.53%	1.6%	0.19	1.49%	290.9

Comparing the outcome of the cooperative case with the noncooperative one, it is realized that the rise in GHGs is smaller, and consequently the increase in the temperature is smaller. With $\mu = 0.1$ GHGs rise by about factor of 1.6, implying an increase in temperature by 2.3 °C. This is due to higher abatement shares in the cooperative world and, as a consequence, smaller emissions in all regions. As for the qualitative results, we see that they do not differ from the last section.

It can also be seen that emissions are clearly smaller than in the noncooperative case. In region 1 emissions are 50 percent smaller, in region 2, 44 percent, and in region 3, there are 37 percent fewer emissions compared to the noncooperative world. The reason emissions in regions 1 and 2 in the cooperative case are much smaller than in the noncooperative case compared with region 3 is that the shadow price of emissions for regions 1 and 2 in the cooperative case is much higher in absolute values than in the noncooperative case. This holds because in the cooperative case regions 1 and 2 take into account not only their own damages but also damages in region 3.

Furthermore, growth rates tend to be larger in the cooperative world. This holds for all three regions and is due to the smaller rise in the average surface temperature. But in region 3, still no positive sustained growth can be observed for $\mu = 0.005$.

From a qualitative point of view the results remain the same for $\mu = 0.005$. In particular, emissions in region 3 are again higher than in region 2. In the noncooperative case the same outcome could be observed. Further, the increase in the average global surface temperature is much larger compared to the case $\mu = 0.1$.

In table 10.7 we study the model assuming that welfare in region 3 gets a weight that is double the weight given to welfare in regions 1 and 2, that is, $w_3 = 2w_1 = 2w_2 = 2$, where we limit our consideration to the case $\mu = 0.1$. A possible justification for higher weights can be seen by thinking of the Rawls criterion, according to which welfare in an economy is determined by the poorest. Then one can argue that welfare in the poorest region should receive a higher weight in the world.[7]

---

[7] Of course, a strict application of the Rawls criterion would require maximizing welfare of only the poorest region, which, however, would not be a cooperative solution.

Table 10.7 shows that now region 3 has a smaller abatement share and higher emissions if its welfare gets a higher weight compared to the case where all three regions get the same weight. The other two regions have higher abatement shares and smaller emissions. As a result, the growth rates in regions 1 and 2 tend to fall while that in region 3 tends to rise.

10.3.3 Equal Marginal Damages in All Regions

In this section we consider the world economy where abatement ratios are set such that marginal damages in steady state are equalized in all regions, given that the steady-state increase in temperature attains an exogenously determined level (marginal damage rule). A justification for considering this scenario is that countries have agreed to limit overall GHG emissions to a certain steady-state level M^*, implying the steady-state temperature T^*. Furthermore, countries are myopic and only consider present damages but not future damages, which are difficult to estimate. Then each country sets the abatement share such that it maximizes its instantaneous return, where the return is the reduction in the damage resulting from the rise in temperature, implying that the marginal benefit of the abatement share equals its marginal cost.

Technically, we proceed as follows. From (10.23) and (10.22) we get in steady state

$$M^* = \frac{\beta_1}{\mu} \sum_{j=1}^{n} \left(\frac{a_j}{\tau_{B,j}} \right)^{\gamma_j}. \tag{10.49}$$

Inserting (10.49) in (8.5) gives the temperature on the BGP as a function of the abatement shares.

Abatement shares, $\tau_{B,j}$, then are obtained as the solution of the equations

$$\dot{T} = 0 \quad \text{and} \quad \frac{\partial D_i}{\partial \tau_{B,i}} = \frac{\partial D_j}{\partial \tau_{B,j}}, \ i \neq j, \tag{10.50}$$

subject to $T^* = \bar{T}$, where $\bar{T}$ is the temperature going along with the exogenously fixed level of GHGs M^*. The balanced growth rate is again given by $(10.21)/K_i$.

To get insight into the model, we compare the outcome of the cooperative world (with equal weights) with that where marginal damages are equalized. Table 10.8 shows the result where T^* is set to $T^* = \bar{T} = 290.7$.

Table 10.8 Abatement Shares, Emissions, and Balanced Growth Rates for the Three Regions as Well as Average Surface Temperature with Equal Marginal Damages (for $\mu = 0.1$).

$\tau^o_{B,1}$	E_1	g_1	$\tau^o_{B,2}$	E_2	g_2	$\tau^o_{B,3}$	E_3	g_3	$T^\star$
0.45%	0.083	2.79%	0.64%	0.12	5.61%	2.2%	0.131	1.33%	290.7

As far as emissions go, one realizes that worldwide emissions do not change,[8] but emissions in the regions are different compared with the cooperative world. So emissions in regions 1 and 2 are 22 and 8.3 percent higher, respectively, and in region 3 they are 15.5 percent lower compared to the cooperative world. This is due to the smaller respectively higher abatement shares in the regions and tends to raise the growth rates in regions 1 and 2 and lower the growth rate in region 3.

The reason the abatement share in region 3 is higher and emissions are lower compared to the cooperative solution is that the direct negative effects of the temperature increase are more important when marginal damages are equated. In the dynamic optimization problem, the damage going along with a temperature increase was taken into account only indirectly, as a side condition, and the goal was to maximize the discounted stream of utility resulting from consumption. In the scenario in this section, equating marginal damages is the goal, and consequently the region with the higher marginal damages, resulting from a temperature rise, has to choose a higher abatement share and emit less.

Thus, our calculations show that in the scenario where the marginal damage rule is applied, emissions in all regions are different compared with the outcome of the cooperative solution. There are changes concerning the level of GHG emissions of up to 20 percent. In general, it can be said that those countries with high marginal damages will emit less than in the cooperative solution.

---

[8] Of course, this is due to the requirement that the temperature increase is the same as in the cooperative world.

11

A Model with Optimizing Agents

Up to now we have considered descriptive models of ongoing growth. Here, we study a model of endogenous growth and climate change assuming optimizing agents. We start with the description of the structure of our economy.

11.1 THE COMPETITIVE ECONOMY

11.1.1 The Structure of the Economy

We consider an economy where one homogeneous good is produced. Furthermore, the economy is represented by one individual with household production who maximizes a discounted stream of utility arising from per capita consumption times the number of household members subject to a budget constraint (see Greiner 2004b). A frequently used standard utility function U in economics is

$$U = \begin{cases} (C D_1 (T - T_o))^{1-\sigma}/(1 - \sigma), & \text{for } \sigma > 0, \sigma \neq 1 \\ \ln C + \ln D_1 (T - T_o), & \text{for } \sigma = 1. \end{cases} \quad (11.1)$$

C denotes per capita consumption and σ is the inverse of the intertemporal elasticity of substitution of consumption between two points in time. $D_1(T - T_o)$ gives the disutility resulting from deviations from the normal temperature T_o.[1] As to the functional form of $D_1(T - T_o)$, we assume that it is given by equation (9.5) with derivative as in (9.6). Note that we now suppose that deviations from the preindustrial surface temperature have direct impacts on the individual's utility. We do this because higher temperatures lead to more extreme weather events, like storms or flooding, which reduce people's well-being.

The individual's budget constraint in per capita terms is given by[2]

$$Y(1 - \tau) = \dot{K} + C + B + \tau_E E L^{-1} + (\delta + n)K, \quad K(0) = K_0, \quad (11.2)$$

with Y per capita production, K per capita capital, B per capita abatement activities, and E emissions. $\tau \in [0, 1)$ is the tax rate on production,

---

[1] The normal temperature is the preindustrial temperature.
[2] The per capita budget constraint is derived from the budget constraint with aggregate variables, denoted by the subscript g, according to $\dot{K}/K = \dot{K_g}/K_g - \dot{L}/L$.

and $\tau_E > 0$ is the tax on emission. δ is the depreciation rate of capital. L is labor, which grows at rate n. In our model formulation abatement is a private good.[3] The production function is given by

$$Y = A K^\alpha \bar{K}^{1-\alpha} D_2(T - T_0), \tag{11.3}$$

with K per capita capital, $\alpha \in (0,1)$ the capital share, and A a positive constant. $D_2(T - T_0)$ is the damage due to deviations from the normal temperature T_o and has the same functional form as $D_1(\cdot)$. We will concretely specify both $D_1(\cdot)$ and $D_2(\cdot)$. $\bar{K}$ gives positive externalities from capital resulting from spillovers. This assumption implies that in equilibrium the private gross marginal returns to capital[4] are constant and equal to $\alpha A D_2(\cdot)$, thus generating sustained per capita growth if A is sufficiently large. This is the simplest endogenous growth model existing in the economics literature. However, because we are interested less in explaining sustained per capita growth but more in the interrelation between global warming and economic growth, this model is sufficiently elaborate.

Concerning emissions of GHGs, we now make a slightly different assumption. We assume that these are a byproduct of capital used in production and expressed in CO_2 equivalents. So emissions are a function of per capita capital relative to per capita abatement activities. This implies that a higher capital stock goes along with higher emissions for a given level of abatement spending. It should also be mentioned that the emission of GHGs does not affect utility and production directly, but only indirectly by affecting the climate of the Earth, which leads to a higher surface temperature and more extreme weather situations. Formally, emissions are described by

$$E = \left(\frac{aK}{B}\right)^\gamma, \tag{11.4}$$

with $\gamma > 0$ and $a > 0$ constants. The parameter a can be interpreted as a technology index describing how polluting a given technology is. For large values of a a given stock of capital (and abatement) goes along with high emissions, implying a relatively polluting technology and vice versa.

The government in our economy is modeled very simply. The government's task is to correct the market failure caused by the negative

---

[3] There exist some contributions that model abatement as a public good. See Ligthart and van der Ploeg (1994) or Nielsen et al. (1995) and part I.

[4] With gross return we mean the return to capital before tax and for the temperature equal to the preindustrial level.

environmental externality.[5] The revenue of the government is used for nonproductive public spending so that it does not affect the consumption-investment decision.

We also point out that we only consider an emission tax and not other environmental policies, such as tradeable permits. We do this because we consider a representative agent. We do not have multiple actors in this chapter who can trade permits. Therefore, we consider the emission tax as the regulatory instrument. However, we are aware that under certain more realistic scenarios, permits may be superior to taxation as an environmental policy measure. Permits might become important, in particular when it is difficult for the government to evaluate marginal costs and benefits of abatement so that the effects of an environmental tax are difficult to evaluate. In this case, permits that limit the quantity of total emissions are preferable.[6]

The individual's optimization problem can be written as

$$\max_{C,A} \int_0^\infty e^{-\rho t} L_0 e^{nt} U(C, D_1(T - T_o)) dt, \tag{11.5}$$

subject to (11.1), (11.2), (11.3), and (11.4). ρ in (11.5) is the subjective discount rate; L_0 is labor supply at time $t = 0$, which we normalize to unity and grows at constant rate n. Note that in the competitive economy the individual takes into account neither the negative externality of capital, the emission of GHGs, nor the positive externalities, that is, the spillover effects.

To find the optimal solution, we form the current-value Hamiltonian, which is

$$\mathcal{H}(\cdot) = (C D_1(T - T_o))^{1-\sigma} / (1 - \sigma) + \lambda_1((1 - \tau) A K^\alpha \bar{K}^{1-\alpha} D_2(T - T_o)$$

$$- C - A - \tau_E L^{-1} a^\gamma K^\gamma B^{-\gamma} - (\delta + n)K), \tag{11.6}$$

with λ_1 the shadow price of K. Note that we used $E = a^\gamma K^\gamma B^{-\gamma}$.

The necessary optimality conditions are given by

$$\frac{\partial \mathcal{H}(\cdot)}{\partial C} = C^{-\sigma} D_1^{1-\sigma} - \lambda_1 = 0, \tag{11.7}$$

$$\frac{\partial \mathcal{H}(\cdot)}{\partial B} = \tau_E L^{-1} a^\gamma K^\gamma \gamma B^{-\gamma-1} - 1 = 0, \tag{11.8}$$

$$\dot{\lambda}_1 = (\rho + \delta)\lambda_1 - \lambda_1 \left((1 - \tau) A \alpha D_2(\cdot) - (\tau_E/LK) \gamma a^\gamma K^\gamma B^{-\gamma}\right). \tag{11.9}$$

---

[5] How the government has to take into account the positive externality is studied in section 5.
[6] For an extensive treatment of permits and their implementation problems when used as regulator instruments to correct for market failure, see Chichilnisky (2004).

In (11.9) equilibrium $K = \bar{K}$ holds. Further, the limiting transversality condition $\lim_{t \to \infty} e^{-(\rho+n)t} \lambda_1 K = 0$ must hold.

Using (11.7) and (11.9) we can derive a differential equation giving the growth rate of per capita consumption. This equation is obtained as

$$\frac{\dot{C}}{C} = -\frac{\rho + \delta}{\sigma} + \frac{\alpha}{\sigma}(1 - \tau) A D_2(\cdot) - \frac{\gamma}{\sigma} \frac{\tau_E}{LK} a^\gamma K^\gamma B^{-\gamma} + \frac{1 - \sigma}{\sigma} \frac{D_1'(\cdot)}{D_1(\cdot)} \dot{T},$$
(11.10)

where $D_1'(\cdot)$ stands for the derivative of $D_1(\cdot)$ with respect to T. Combining (11.8) and (11.4) yields

$$E = \left(\frac{\tau_E}{LK}\right)^{-\gamma/(1+\gamma)} a^{\gamma/(1+\gamma)} \gamma^{-\gamma/(1+\gamma)}.$$
(11.11)

Using (8.5) and (8.6) from chapter 8 with the numerical parameter values introduced and the equations derived in this section, the competitive economy is completely described by the following differential equations

$$\dot{T} c_h = 71.7935 - 5.6710^{-8}(19.95/109)T^4 + 6.3\beta_1 (1 - \xi) \ln \frac{M}{M_o},$$

$$T(0) = T_0 \tag{11.12}$$

$$\dot{M} = \beta_2 \left(\frac{\tau_E}{LK}\right)^{-\gamma/(1+\gamma)} a^{\gamma/(1+\gamma)} \gamma^{-\gamma/(1+\gamma)} \mu M,$$

$$- M(0) = M_0, \tag{11.13}$$

$$\frac{\dot{C}}{C} = -\frac{\rho + \delta}{\sigma} + \frac{\alpha}{\sigma}(1 - \tau) A D_2(\cdot) - \frac{\gamma}{\sigma} \left(\frac{\tau_E}{LK}\right)^{1/(1+\gamma)}$$

$$\times a^{\gamma/(1+\gamma)} \gamma^{-\gamma/(1+\gamma)} + \frac{1 - \sigma}{\sigma} \frac{D_1'(\cdot)}{D_1(\cdot)} \dot{T}, \tag{11.14}$$

$$\frac{\dot{K}}{K} = A D_2(T - T_0)(1 - \tau) - \left(\frac{\tau_E}{LK}\right)^{1/(1+\gamma)} a^{\gamma/(1+\gamma)} \gamma^{-\gamma/(1+\gamma)}(1 + \gamma)$$

$$- \frac{C}{K} - (\delta + n), \quad K(0) = K_0, \tag{11.15}$$

where $C(0)$ can be chosen by society.

11.1.2 Analytical Results

Before we study comparative static properties of our model on the BGP, we analyze its dynamics. Again, a BGP is defined as a path on which variables grow at the same rate and GHGs as well as the average surface temperature are constant. The next proposition shows that there exists a unique BGP for this economy under a slight additional assumption.

Proposition 19 *For the competitive model economy there exists a unique BGP for a constant value of τ_E/LK that is a saddle point with one positive and two negative real eigenvalues.*

Proof: To show uniqueness of the steady state we solve (11.12) = 0, (11.13) = 0, and (11.14) = (11.15) with respect to T, M, and $c \equiv C/K$. Setting (11.12) = 0 gives

$$T_{1,2} = \pm 99.0775 \, (71.7935 + 6.3 \, \beta_1 \, (1 - \xi) \, \ln(M/M_0))^{1/4}$$

$$T_{3,4} = \pm 99.0775 \sqrt{-1} \, (71.7935 + 6.3 \, \beta_1 \, (1 - \xi) \, \ln(M/M_0))^{1/4}.$$

Clearly $T_{3,4}$ are not feasible. Further, because $M \geq M_0$ only the positive solution of $T_{1,2}$ is feasible. Uniqueness of M and c on the BGP is immediately seen.[7] To study the local dynamics we note that the economy around the BGP is described by (11.12), (11.13), and $\dot{c} = c \, (\dot{C}/C - \dot{K}/K)$. The Jacobian matrix J corresponding to this dynamic system is obtained as

$$J = \begin{pmatrix} -4 \, (5.67 \, 10^{-8}) \, (19.95/109)(T^\star)^3 \, c_h^{-1} & 6.3 \, \beta_1 \, (1 - \xi) \, (M^\star)^{-1} c_h^{-1} & 0 \\ 0 & -\mu & 0 \\ a_{31} & \frac{1-\sigma}{\sigma} \frac{D_1'(\cdot)}{D_1} c^\star 6.3 \, \beta_1 \, (1 - \xi)(M^\star)^{-1} c_h^{-1} & c^\star \end{pmatrix},$$

with $\star$ denoting steady-state values and

$$a_{31} = c^\star \left((1 - \tau) \, A \, D_2'(\cdot) \left(\frac{\alpha}{\sigma} - 1 \right) + \frac{1 - \sigma}{\sigma} \right.$$
$$\left. \times \left(\frac{D_1''(\cdot) \, D_1(\cdot) - D_1'(\cdot)^2}{D_1(\cdot)^2} \, \dot{T} + \frac{\partial \dot{T}}{\partial T} \frac{D_1'(\cdot)}{D_1(\cdot)} \right) \right).$$

The eigenvalues of J are given by

$$e_1 = -4(5.67 \, 10^{-8})(19.95/109) \, (T^\star)^3 \, c_h^{-1}, \quad e_2 = -\mu \quad \text{and} \quad e_3 = c^\star.$$

Thus the proposition is proved.   □

Proposition 19 shows that there exists a two-dimensional stable manifold. Solutions starting on that manifold converge to the BGP in the

---

[7] We neglect the economically meaningless steady state $c^\star = 0$.

long run while all other solutions diverge. Since $T(0)$, $M(0)$, and $K(0)$ are given, the value for $C(0)$ must be chosen such that $c(0) \equiv C(0)/K(0)$ lies on the stable manifold. Further, it cannot be excluded that the BGP goes along with a negative growth rate because the sign of the balanced growth rate depends on the concrete numerical values of the parameters. The question of whether there exists a BGP with a positive growth rate, that is, a nondegenerate BGP, for a certain parameter constellation is addressed in the next section. Here we make the assumption that a nondegenerate BGP exists.

An important aspect is that on a BGP M is constant, implying that emissions of GHGs are constant, too. In a growing economy, however, this is only possible if abatement activities rise with the same rate as the capital stock. Because abatement activities are set by private agents to satisfy (11.8), the government has to raise the emission tax in such a way that the ratio τ_E/LK is constant. This implies that the tax on the emission must rise with the same rate as the aggregate capital stock LK. However, note that this only holds for a given relation between the ratio K/B and emissions. So technical progress, generating a less polluting technology, which implies that a given capital stock K causes less emissions, would change this outcome. In our framework this would be modeled by a decrease in a. This would affect the value of τ_E. In the following we do not go into the details of this aspect but analyze our model for constant parameter values and assume that the government keeps the ratio τ_E/LK constant.

In a next step we analyze how the balanced growth rate, which we denote by g and which is given by (11.14) with $\dot{T} = 0$, reacts to changes in the production tax rate τ and to different values of the ratio τ_E/LK. To do so, we differentiate g with respect to τ. This gives

$$\frac{\partial g}{\partial \tau} = A \left(\frac{\alpha}{\sigma} \right) \left((1 - \tau) D_2'(\cdot) \frac{\partial T^\star}{\partial \tau} - D_2(\cdot) \right),$$

with $T^\star$ denoting the surface temperature on the BGP. From (11.12) and (11.13) it is immediately seen that $\partial T^\star/\partial \tau = 0$, so that a higher tax rate on production unequivocally lowers the balanced growth rate.

This does not hold for variations of the emission tax. Differentiating g with respect to (τ_E/LK) yields

$$\frac{\partial g}{\partial(\tau_E/LK)} = A \left(\frac{\alpha}{\sigma} \right) (1 - \tau) D_2'(\cdot) \frac{\partial T^\star}{\partial(\tau_E/LK)} - \frac{\gamma}{\sigma(1 + \gamma)} E, \qquad (11.16)$$

with E given by (11.11). To get an idea about the sign of this expression, we need to know the sign of $\partial T^\star/\partial(\tau_E/LK)$. $\partial T^\star/\partial(\tau_E/LK)$ is obtained by solving (11.12) with respect to $T^\star$, inserting $M^\star$ obtained from (11.13) $= 0$, and then differentiating with respect to (τ_E/LK). It is immediately seen that this derivative is negative. Because $T \geq T_o$, which holds on a BGP due to $M \geq M_o$, the derivative of $D_2(\cdot)$ is negative so that the first part of the foregoing expression is positive.

With these considerations we can state that this derivative shows that an increase in the emission tax may raise or lower the balanced growth rate. This holds because there are two counteracting forces: on one hand, a higher emission tax reduces the net marginal product of private capital, thus reducing the balanced growth rate. On the other hand, a higher emission tax reduces the average surface temperature and, as a consequence, the damage resulting from deviations of the actual temperature from its preindustrial level. This tends to raise the net marginal product of private capital and the incentive to invest. So we can state that a higher tax on GHG emissions may yield a win-win situation, or a double dividend, by both raising the balanced growth rate and reducing GHG emissions. It should also be mentioned that the direct negative effect of global warming as to utility does not affect this outcome. This is easily seen from (11.11) and (11.12) with M^* determined by (11.13) = 0.

These considerations suggest that there exists a growth-maximizing value for the emission tax ratio. This implies that a so-called double dividend may exist: a higher emissions tax leads to both less emissions and higher growth and, as a consequence, to higher welfare.

This general result is not too surprising because our model can be seen as a special variant of a more general model where emissions raise a stock of pollutants, which negatively affects utility and productivity in an economy. For this general class it is known that a double dividend may exist. However, our goal is not only to state that such an outcome is feasible but to derive more concrete results for the problem of global warming where insights from physics are used. We intend to analyze our model for parameter values that are considered reasonable and study growth and welfare effects of different emissions tax ratios.

To study these questions, it is necessary to concretely specify the two damage functions $D_1(T - T_o)$ and $D_2(T - T_o)$ and resort to numerical examples. This is done next.

NUMERICAL EXAMPLES

We analyze how the growth rate and welfare react to variations in the emission tax ratio. We do this for the economy on the BGP using numerical examples, and we start this section with a description of the parameter values employed in our numerical analysis.

We consider one time period to comprise one year. The discount rate is set to $\rho = 0.03$, the population growth rate is assumed to be $n = 0.02$, and the depreciation rate of capital is $\delta = 0.075$. The preindustrial level of GHGs is normalized to one, $M_o = 1$. γ, β_1, and ξ are set to $\gamma = 1$, $\beta_1 = 1.1$, and $\xi = 0.3$ (for the latter two, see the discussion in chapter 8). The tax rate on output is $\tau = 0.1$, and the capital share is $\alpha = 0.45$. This value seems high. However, if capital is considered in a broad sense—meaning that it also comprises human capital—this value is reasonable.[8] A is set to $A = 0.35$, implying that the social gross marginal return to capital is 35 percent for $T = T_o$.

As for τ_E/LK, we consider the values $\tau_E/LK = 0.005, 0.01, 0.015$. For example, in Germany the ratio of tax on mineral oil to private gross capital (excluding residential capital) was 0.0037 (0.0068) in 1999 (see Statistisches Bundesamt 2000, pp. 510, 639). a is set to $a = 1.65\,10^{-3}$ so that a doubling of GHGs implies a rise in the average surface temperature of about 3.3 °C for $\tau_E/LK = 0.01$. We also consider the lower values $\tau_E/LK = 0.0005, 0.001, 0.0015$.[9] As for the inverse of the intertemporal elasticity of substitution of consumption, we consider the values $\sigma = 1.1$ and $\sigma = 2$.

An important role is played by the damage functions $D_1(\cdot)$ and $D_2(\cdot)$. These are the same as in equation (9.5). For $D_1(\cdot)$ we assume the function

$$D_1(\cdot) = \left(a_1\,(T - T_o)^2 + 1\right)^{-\varphi}, \tag{11.17}$$

with $a_1 > 0$, $\varphi > 0$. $D_2(\cdot)$ is given by

$$D_2(\cdot) = \left(a_2\,(T - T_o)^2 + 1\right)^{-\phi}, \tag{11.18}$$

with $a_2 > 0$, $\phi > 0$.

For the numerical values of the parameters in (11.17) we assume $a_1 = 0.035$ and $\varphi = 0.035$, which are left unchanged in the examples except for table 11.4. These values imply that a rise of the surface temperature by 1 (2, 3) degree(s) implies a decrease of utility by 0.012 (0.046, 0.096) percent for $\sigma = 1.1$. For $\sigma = 2$ the decrease is 0.12 (0.46, 0.96) percent for a temperature increase of 1 (2, 3) degree(s). Setting $a_1 = 0.07$ and $\varphi = 0.07$ gives a decrease in utility of 0.047, 0.17, 0.34 (0.47, 1.74, 3.48) percent for a temperature increase of 1, 2, 3 degrees with $\sigma = 1.1$ ($\sigma = 2$).

In IPCC (1996), pp. 196–197, an example is given how a monetary value can be attached to a change in the risk of death as a result of a climate change. The IPCC cites the study by Fankhauser (1995) who estimates this number to be on average 0.26 percent of world GDP for a doubling of CO_2 concentration,[10] which corresponds in our model to a temperature increase of about 3.3 °C. Note that there are other effects that have an impact on individual's utility, like an increase in extreme

---

[8] The choice of α does not affect the qualitative results, only the magnitude of endogenous variables like the balanced growth rate for example.
[9] To get temperature increases for a doubling of GHGs in line with IPCC calculations, we set $a = 1.65\,10^{-4}$ in this case.
[10] This holds for the economy in steady state.

Table 11.1 Varying the Emission Tax between 0.005 and 0.015 with $m_2 = 0.025, \phi = 0.025$.

τ_E/LK	σ	$T^\star$	$M^\star$	g	W	$B^\star/K^\star$
0.005	1.1	293.3	2.8	0.0293	-388.612	$2.87 \, 10^{-3}$
0.005	2	293.3	2.8	0.0161	$-3.659 \, 10^{-2}$	$2.87 \, 10^{-3}$
0.01	1.1	291.7	1.99	0.0289	-389.291	$4.06 \, 10^{-3}$
0.01	2	291.7	1.99	0.0159	$-3.656 \, 10^{-2}$	$4.06 \, 10^{-3}$
0.015	1.1	290.7	1.6	0.0285	-390.615	$4.97 \, 10^{-3}$
0.015	2	290.7	1.6	0.0157	$-3.686 \, 10^{-2}$	$4.97 \, 10^{-3}$

weather events, for example, so that different values may be plausible, too. Furthermore, it should also be kept in mind that empirical estimates for damages are necessarily uncertain, as already mentioned. This holds in particular for the direct effect of global warming concerning people's utility.

As far as the numerical values of a_2 and ϕ, we consider the cases $(a_2 = 0.025, \phi = 0.025)$, $(a_2 = 0.035, \phi = 0.035)$, and $(a_2 = 0.05, \phi = 0.05)$. The combination $(a_2 = 0.025, \phi = 0.025)$ implies that an increase of the surface temperature by 1 (2, 3) degree(s) leads to a decrease of aggregate production by 0.06 (0.2, 0.5) percent. With $(a_2 = 0.05, \phi = 0.05)$ an increase of the surface temperature by 1 (2, 3) degree(s) leads to a decrease of aggregate production by 0.2 (0.9, 1.8) percent. Comparing these values with the estimates published in IPCC (1996), we see that most of the values we choose tend to be within the range of that study.

In the following tables we report the results of our numerical studies. These studies compare different scenarios where it is assumed that the exogenous variables, that is, τ_E/LK and σ, take their respective values for all $t \in [0, \infty)$ and the economy immediately jumps to its BGP. Table 11.1 shows the effects of different ratios τ_E/LK where the * denotes values on the BGP. g is the balanced growth rate given by (11.14) with $\dot{T} = 0$ and W denotes welfare. W is given by

$$W = \int_0^\infty e^{-(\rho - n)t} U(C(t), D_1(T^\star - T_o)) dt,$$

with $U(\cdot)$ given by (11.1) and $C(t) = C^\star e^{gt}$. $C^\star$ is determined endogenously by (11.15) such that $\dot{K}/K = g$ holds for all $t \in [0, \infty)$. As to $K_0 = K^\star$ we set $K_0 = 5500$, which is about the capital stock in Germany (in billion euros) in 2000 (see Institut der Deutschen Wirtschaft 2003).[11]

---

[11] The numerical value of K_0 does not affect the qualitative outcome, only the absolute value of welfare W.

Table 11.2 Varying the Emission Tax between 0.005 and 0.015 with $m_2 = 0.05, \phi = 0.05$.

τ_E/LK	σ	$T^\star$	$M^\star$	g	W	$B^\star/K^\star$
0.005	1.1	293.3	2.8	0.0258	-399.396	$2.87\ 10^{-3}$
0.005	2	293.3	2.8	0.0142	$-3.95\ 10^{-2}$	$2.87\ 10^{-3}$
0.01	1.1	291.7	1.99	0.0269	-395.408	$4.06\ 10^{-3}$
0.01	2	291.7	1.99	0.0148	$-3.817\ 10^{-2}$	$4.06\ 10^{-3}$
0.015	1.1	290.7	1.6	0.0273	-394.099	$4.97\ 10^{-3}$
0.015	2	290.7	1.6	0.015	$-3.778\ 10^{-2}$	$4.97\ 10^{-3}$

We also report the steady-state ratio of abatement activities to capital, $B^\star/K^\star$.

Table 11.1 shows that an increase in the emission tax reduces the balanced growth rate independent of the value of σ. However, the value of σ affects welfare. For $\sigma = 1.1$ welfare decreases as a consequence of the lower balanced growth rate. For $\sigma = 2$, welfare first rises with an increase in the emission tax rate, although the balanced growth rate declines. In this case, the negative effect of a lower balanced growth rate on welfare is dominated by the positive direct welfare effect resulting from a lower increase in the average surface temperature. If the emission tax rate is further increased, welfare declines again. The effect that an increase in the emission tax rate lowers the balanced growth rate but raises welfare is more likely the larger the direct effect of a temperature increase on utility. This is demonstrated in table 11.4 where we set $m_1 = 0.07$ and $\varphi = 0.07$.

In case of $\sigma = 1.1$, maximum welfare is obtained for about $\tau_E/LK = 0.001$. However, this implies that the temperature increase is approximately $8\,°C$. Here it must be mentioned that the assumption of a continuous damage function is only justified provided the rise in temperature does not exceed a certain threshold. This holds because for higher increases of the temperature catastrophic events may occur, going along with extremely high economic costs that cannot be evaluated. Just one example is the breakdown of the Gulf Stream, which would dramatically change the climate in Europe. Therefore, the analysis assuming a damage function only makes sense for temperature increases within certain bounds. In table 11.2 we consider our model on the BGP for the case ($m_2 = 0.05, \phi = 0.05$).

Table 11.2 shows that an increase in the damage caused by a higher average surface temperature yields a double dividend.[12] In this case, a rise in the emission tax yields both a higher balanced growth rate and a

---

[12] It should be recalled that this effect is independent of the function $D_1(\cdot)$, which determines the direct utility effect of global warming.

Table 11.3 Varying the Emission Tax between 0.0005 and 0.0015 with $a_2 = 0.025, \phi = 0.025$.

τ_E/LK	σ	$T^\star$	$M^\star$	g	W	$B^\star/K^\star$
0.0005	1.1	293.3	2.8	0.0316	-381.226	$2.87\ 10^{-4}$
0.0005	2	293.3	2.8	0.0174	$-3.435\ 10^{-2}$	$2.87\ 10^{-4}$
0.001	1.1	291.7	1.99	0.0323	-378.88	$4.06\ 10^{-4}$
0.001	2	291.7	1.99	0.0177	$-3.343\ 10^{-2}$	$4.06\ 10^{-4}$
0.0015	1.1	290.7	1.6	0.0325	-377.85	$4.97\ 10^{-4}$
0.0015	2	290.7	1.6	0.0179	$-3.303\ 10^{-2}$	$4.97\ 10^{-4}$

lower rise in the average global surface temperature independent of the intertemporal elasticity of substitution of consumption. Of course, this implies that welfare along the BGP unequivocally rises. The maximum growth rate and maximum welfare are obtained for about $\tau_E/LK = 0.02$, implying $M^\star = 1.4$ and $T^\star = 290.1$.

We do not show a table for the case ($a_2 = 0.035$, $\phi = 0.035$). The results for this case are in between the results obtained in tables 11.1 and 11.2. This means that rising τ_E/LK from 0.005 to 0.01 first raises the balanced growth rate. Increasing τ_E/LK further from 0.01 to 0.015 then reduces the balanced growth rate, which, however, remains larger than for $\tau_E/LK = 0.005$. This holds for both $\sigma = 1.1$ and $\sigma = 2$. The same holds as concerns welfare, that is, welfare first rises with an increase in τ_E/LK and then declines but remains higher than for the initial level of τ_E/LK.

Next we consider the case of smaller values for the emission tax ratio. The results for ($a_2 = 0.025$, $\phi = 0.025$) are shown in table 11.3.

Table 11.3 shows that with smaller values for the emission tax ratio, a double dividend is obtained even in the case where the damage caused by a climate change is relatively small. Thus, the smaller the emission tax ratio, the more likely it is that raising this policy variable both raises the long-run growth rate and reduces the increase in the average surface temperature. The reason for the outcome in table 11.3 is that the negative growth effect of an increase in the emission tax is smaller if the technology in use is less polluting. This is seen from (11.16) together with (11.11). From an economic point of view, the interpretation is as follows.

If the emission tax rate is increased, the individual shifts resources from investment to abatement. The rise in abatement relative to the capital stock to get a certain decrease in emissions, however, is smaller when the technology in use is relatively clean.[13] Consequently, the

---

[13] This is seen by differentiating B/K, obtained from (11.8), with respect to τ_E/LK.

Table 11.4 Varying the Emission Tax between 0.005 and
0.015 with $m_2 = 0.025, \phi = 0.025$ and
$m_1 = 0.07, \varphi = 0.07$.

τ_E/LK	σ	T^*	M^*	g	W	A^*/K^*
0.005	1.1	293.3	2.8	0.0322	-390.476	$2.87\ 10^{-3}$
0.005	2	293.3	2.8	0.0161	$-3.839\ 10^{-2}$	$2.87\ 10^{-3}$
0.01	1.1	293.3	1.99	0.0318	-390.395	$4.06\ 10^{-3}$
0.01	2	291.7	1.99	0.0159	$-3.761\ 10^{-2}$	$4.06\ 10^{-3}$
0.015	1.1	290.7	1.6	0.0313	-391.262	$4.97\ 10^{-3}$
0.015	2	290.7	1.6	0.0157	$-3.748\ 10^{-2}$	$4.97\ 10^{-3}$

negative growth effect is smaller compared to a situation with a more polluting technology. The balanced growth rate, is maximized for about $\tau_E/LK = 0.003$. Of course, the double dividend is also obtained for $(a_2 = 0.035, \phi = 0.035)$ and $(a_2 = 0.05, \phi = 0.05)$. In the latter cases, the effects are larger in magnitude compared to those in table 11.3.

In table 11.4 we show that with a higher direct damage of a rise in temperature concerning utility, an increase in the emission tax reduces the balanced growth rate but leads to higher welfare. This holds for $\sigma = 2$ but for $\sigma = 1.1$ only when τ_E/LK is increased from 0.005 to 0.01. In the latter case, raising τ_E/LK further reduces welfare. Together with table 11.1, this shows that a rise in welfare going along with a smaller growth rate is the more likely the smaller the intertemporal elasticity of substitution of consumption $1/\sigma$. The reason for that outcome is that with a smaller intertemporal elasticity of substitution the individual is less willing to shift utility benefits into the future. Therefore, he prefers a higher utility today, resulting from a smaller temperature increase, to a higher growth rate of consumption, which would yield utility only in the future.

In next section we present and analyze the social optimum and compare the outcome with the competitive model economy.

11.2 THE SOCIAL OPTIMUM

In formulating the optimization problem, a social planner takes into account both the positive and negative externalities of capital. Consequently, for the social planner the resource constraint is given by

$$\dot{K} = AK D_2(T - T_o) - C - B - (\delta + n)K, K(0) = K_0. \tag{11.19}$$

The optimization problem is[14]

$$\max_{C,A} \int_0^\infty e^{-\rho t} L_0 e^{nt} \left(\ln C + \ln D_1(T - T_o) \right) dt, \qquad (11.20)$$

subject to (11.19), (8.5), (8.6), and (11.4), where $D_1(\cdot)$ and $D_2(\cdot)$ are again given by (11.17) and (11.18).

To find necessary optimality conditions we formulate the current-value Hamiltonian, which is

$$\mathcal{H}(\cdot) = \left(\ln C + \ln D_1(T - T_o) \right) + \lambda_2 (A\,KD_2(T - T_o) - C - B - (\delta + n)K)$$

$$+ \lambda_3 \left(\beta_2\, a^\gamma K^\gamma B^{-\gamma} - \mu M \right) + \lambda_4 \, (c_h)^{-1} \cdot \left(\frac{1367.5}{4} \, 0.21 \right.$$

$$\left. - (5.67\,10^{-8})(19.95/109)T^4 + \beta_1\,(1 - \xi)\,6.3\ln\frac{M}{M_o} \right), \qquad (11.21)$$

with λ_i, $i = 2, 3, 4$, the shadow prices of K, M, and T respectively and $E = a^\gamma K^\gamma A^{-\gamma}$. Note that λ_2 is positive while λ_3 and λ_4 are negative.

The necessary optimality conditions are obtained as

$$\frac{\partial \mathcal{H}(\cdot)}{\partial C} = C^{-1} - \lambda_2 = 0, \qquad (11.22)$$

$$\frac{\partial \mathcal{H}(\cdot)}{\partial B} = -\lambda_3\,\beta_2\,a^\gamma\,K^\gamma\,\gamma B^{-\gamma-1} - \lambda_2 = 0, \qquad (11.23)$$

$$\dot{\lambda}_2 = (\rho + \delta)\,\lambda_2 - \lambda_2\,A\,D_2(\cdot) - \lambda_3\,\beta_2\,\gamma\,a^\gamma\,K^{\gamma-1}\,B^{-\gamma}, \qquad (11.24)$$

$$\dot{\lambda}_3 = (\rho - n)\,\lambda_3 + \lambda_3\,\mu - \lambda_4\,(1 - \xi)\,\beta_1\,6.3\,c_h^{-1}\,M^{-1}, \qquad (11.25)$$

$$\dot{\lambda}_4 = (\rho - n)\,\lambda_4 - \frac{D_1'(\cdot)}{D_1} - \lambda_2 A K D_2'(\cdot) + \frac{\lambda_4\left((19.95/109)\,4\,T^3\right)}{(5.67\,10^8)\,c_h}. \qquad (11.26)$$

Further, the limiting transversality condition $\lim_{t \to \infty} e^{-(\rho+n)t}(\lambda_2 K + \lambda_3 T + \lambda_4 M) = 0$ must hold.

Comparing the optimality conditions of the competitive economy with that of the social planner demonstrates how the government has to set taxes to replicate the social optimum. The next proposition gives the result.

---

[14] For the social optimum we only study the case $\sigma = 1$.

Proposition 20 *The competitive model economy replicates the social optimum if τ_E/LK and τ are set according to*

$$\frac{\tau_E}{LK} = \beta_2 \frac{-\lambda_3}{\lambda_2 K}, \quad \tau = 1 - \alpha^{-1}.$$

Proof: The first condition is obtained by setting (11.8) = (11.23). The second is obtained by setting the growth rate of per capita consumption in the competitive economy equal to that of the social optimum.   □

This proposition shows that the emission tax per aggregate capital has to be set such that it equals the effective price of emissions, $-\lambda_3\beta_2$, divided by the shadow price of capital times per capita capital, $\lambda_2 K$, for all $t \in [0, \infty)$. This makes the representative household internalize the negative externality associated with capital. Furthermore, it can be seen that as usual, the government has to pay an investment subsidy (or negative production tax) of $\tau = 1 - \alpha^{-1}$. The latter is to make the representative individual take into account the positive spillover effects of capital. The subsidy is financed by the revenue of the emission tax and/or by a nondistortionary tax, like a consumption tax, or a lump-sum tax.

From (11.22) and (11.23) we get

$$\frac{B}{K} = \left(c\,(-\lambda_3)\,\beta_2\,\gamma\,a^\gamma\right)^{1/(1+\gamma)}, \tag{11.27}$$

with $c \equiv C/K$. Using (11.27), (11.22), and (11.24), the social optimum is completely described by the following system of autonomous differential equations

$$\dot{C} = C\left(A D_2(\cdot) - (\rho + \delta) - \left((C/K)\,(-\lambda_3)\,\beta_2\,\gamma\,a^\gamma\right)^{1/(1+\gamma)}\right), \tag{11.28}$$

$$\dot{K} = K\left(A D_2(\cdot) - \frac{C}{K} - \left((C/K)\,(-\lambda_3)\,\beta_2\,\gamma\,a^\gamma\right)^{1/(1+\gamma)} - (\delta + n)\right), \tag{11.29}$$

$$\dot{M} = (C/K)^{-\gamma/(1+\gamma)}\,(-\lambda_3)^{-\gamma/(1+\gamma)}\,\beta_2^{1/(1+\gamma)}\,\gamma^{-\gamma/(1+\gamma)}\,a^{\gamma/(1+\gamma)} - \mu M, \tag{11.30}$$

$$\dot{T} = c_h^{-1}\left(71.7935 - 5.67\,10^{-8}(19.95/109)T^4 + 6.3\beta_1\,(1 - \xi)\ln\frac{M}{M_o}\right), \tag{11.31}$$

$$\dot{\lambda}_3 = (\rho - n)\,\lambda_3 + \lambda_3\,\mu - \lambda_4\,(1 - \xi)\,\beta_1\,6.3\,c_h^{-1}\,M^{-1}, \tag{11.32}$$

$$\dot{\lambda}_4 = (\rho - n)\lambda_4 - \frac{D_1'(\cdot)}{D_1} - A\frac{K}{C}D_2'(\cdot) + \lambda_4\left(5.67\,10^{-8}(19.95/109)c_h^{-1}4T^3\right). \tag{11.33}$$

with $K(0) = K_0$, $M(0) = M_0$, and $T(0) = T_0$.

As for the competitive economy, a BGP is given for variables $T^\star$, $M^\star$, $\lambda_3^\star$, $\lambda_4^\star$, and $c^\star$ such that $\dot{T} = \dot{M} = 0$ and $\dot{C}/C = \dot{K}/K$ holds, with $M \geq M_0$. It should be noted that $\dot{T} = \dot{M} = 0$ implies $\dot{\lambda}_3 = \dot{\lambda}_4 = 0$. Proposition 21 gives sufficient conditions for a unique BGP to exist in the social optimum.

Proposition 21 *Assume that $D_i''(\cdot) < 0$, $i = 1, 2$, and $\gamma < (4/3)\,\kappa(1)\,T_o^4$, with $\kappa(1) = 5.67\,10^{-8}(19.95/109)/(6.3\beta_1\,(1 - \xi))$. Then there exists a unique BGP for the social optimum.*

Proof: To prove the proposition we compute $M^\star$ from (11.30) $= 0$ as

$$M^\star = (-\lambda_3)^{-\gamma/(1+\gamma)}\,(c^\star)^{-\gamma/(1+\gamma)}\,\beta_2^{1/(1+\gamma)}\,\gamma^{-\gamma/(1+\gamma)}\,a^{\gamma/(1+\gamma)}/\mu,$$

with $c^\star = \rho - n$ from $\dot{C}/C = \dot{K}/K$. Inserting $M^\star$ in $\dot{\lambda}_3$ and solving $\dot{\lambda}_3 = 0$ with respect to λ_4 yields

$$\lambda_4^\star = -\kappa(2)(-\lambda_3)^{1/(1+\gamma)},$$

with

$$\kappa(2) = \frac{\rho - n + \mu}{(1 - \xi)\,\beta_1\,6.3\,c_h^{-1}\,(c^\star)^{\gamma/(1+\gamma)}\,\beta_2^{-1/(1+\gamma)}\,\gamma^{\gamma/(1+\gamma)}\,a^{-\gamma/(1+\gamma)}\,\mu} > 0.$$

Inserting $\lambda_4^\star$ and $M^\star$ in $\dot{\lambda}_4$ and $\dot{T}$, respectively, and setting the latter two equations equal to zero leads to

$$(-\lambda_3)^{1/(1+\gamma)} = \frac{-D_1'/D_1 - D_2'\,A/c}{\kappa(3) + \kappa(4)\,T^3}, \tag{11.34}$$

$$(-\lambda_3)^{1/(1+\gamma)} = \left(\frac{\kappa(5)}{e^{(-71.7935 + 5.67\,10^{-8}(19.95/109)T^4)/(6.3\,\beta_1\,(1-\xi))}}\right)^{1/\gamma}, \tag{11.35}$$

with

$$\kappa(3) = (\rho - n)\,\kappa(2) > 0,$$

$$\kappa(4) = 5.67\,10^{-8}(19.95/109)\,c_h^{-1}\,4\,\kappa(2) > 0,$$

$$\kappa(5) = c^{-\gamma/(1+\gamma)}\,\beta_2^{1/(1+\gamma)}\,\gamma^{-\gamma/(1+\gamma)}\,a^{\gamma/(1+\gamma)}/(\mu\,M_0) > 0.$$

Setting the right-hand side in (11.34) equal to the right-hand side in (11.35) gives

$$\frac{-D_1'/D_1 - D_2'\,A/c}{\kappa(5)^{1/\gamma}} = \frac{\kappa(3) + \kappa(4)\,T^3}{e^{(\kappa(6) + \kappa(1)T^4)/\gamma}}, \tag{11.36}$$

with

$$\kappa(6) = -71.7935/(6.3\,\beta_1\,(1-\xi)),$$

$$\kappa(1) = 5.67\,10^{-8}(19.95/109)/(6.3\,\beta_1\,(1-\xi)).$$

For $T = T_o$ the right-hand side of (11.36) is zero and the left-hand side is strictly positive. Further, the derivative of the left-hand side of (11.36) is strictly positive for $D_i''(\cdot) < 0$, $i = 1,2$. The derivative of the right-hand side is given by

$$e^{(-\kappa(6)-\kappa(1)T^4)/\gamma}\,3\,T^2\,\kappa(4)\left(1 - \frac{4\,\kappa(3)\,\kappa(1)\,T}{3\,\gamma\,\kappa(4)} - \frac{4}{3\,\gamma}\,\kappa(1)\,T^4\right).$$

Using $\kappa(1) = 5.67\,10^{-8}(19.95/109)/(6.3\,\beta_1\,(1-\xi))$ shows that $\gamma < (4/3)\,\kappa(1)\,T_o^4$ is a sufficient condition for the derivative to be negative for $T \geq T_o$. Thus, the proposition is proved.   □

Note that this proposition gives conditions that are sufficient but not necessary for a unique BGP so that a unique BGP may exist even if they are not fulfilled. The first condition states that the damage functions are strictly concave.[15] The second gives a condition as to the structural parameter γ, which determines the level of emissions in the economy. Inserting $\beta_1 = 1.1$, $\xi = 0.3$ (see chapter 8), and $T_o = 288.4$ gives $\gamma < 19.73$, which does not impose a severe limitation. For our numerical examples the existence of a unique BGP is always assured in the social optimum.

As for the local dynamics, it is difficult to derive concrete results for the analytical model. This is due, among other things, to the determinant of the Jacobian matrix at the steady state, which may be positive or negative. Nevertheless, something can also be said with respect to the analytical model. This is the contents of proposition 22.

Proposition 22 *Assume there exists a unique BGP for the social optimum. Then there exists at most a two-dimensional stable manifold. Further, a Hopf bifurcation can be excluded.*

Proof: To prove the proposition we first recall that the dynamics around the BGP are described by the dynamic system consisting of the equations $\dot{c}/c = \dot{C}/C - \dot{K}/K = n - \rho + c$, (11.30), (11.31), (11.32), and (11.33). Because $c^\star = 0$ can be excluded because c is raised to a negative power in (11.30), we consider the differential equation for c in the rate of growth. Equation $\dot{c}/c$ shows that $C(0)$ must be chosen such that c takes its steady-state value at $t = 0$, meaning that c is a constant. The dynamics of the

---

[15] This implies that the damage caused by a higher temperature is a convex function of the temperature.

other variables then are described by (11.30), (11.31), (11.32), and (11.33). The Jacobian matrix J is given by

$$
J = \begin{pmatrix}
-\mu & 0 & \dfrac{a^{\gamma/(1+\gamma)}\,\beta_2^{1/(1+\gamma)}\,\gamma^{1-\gamma/(1+\gamma)}\,\left(-\lambda_3{}^\star\right)^{-\gamma/(1+\gamma)}}{\left(-\lambda_3{}^\star\right)(c^\star)^{\gamma}/(1+\gamma)\,(1+\gamma)} & 0 \\[2em]
\dfrac{6.3\,\beta_1\,(1-\xi)}{c_h\,M^\star} & \dfrac{5.67\,10^{-8}\cdot 19.95\cdot 4\,(T^\star)^3}{(-1)\,109\,c_h} & 0 & 0 \\[2em]
\dfrac{6.3\,\beta_1\,(1-\xi)\,\lambda_4^\star}{c_h\,(M^\star)^2} & 0 & \mu - n + \rho & \dfrac{-6.3\,\beta_1\,(1-\xi)}{c_h\,M^\star} \\[2em]
0 & a_{42} & 0 & \begin{array}{l} -n + \rho + \\ \dfrac{5.67\,10^{-8}\cdot 19.95\cdot 4\,(T^\star)^3}{109\,c_h} \end{array}
\end{pmatrix},
$$

with

$$
\begin{aligned}
a_{42} = {}& \frac{12\cdot 5.67\,10^{-8}\cdot 19.95\,\lambda_4^\star\,(T^\star)^2}{109\,c_h} + \frac{2\,a_1\,\varphi}{1 + a_1\,(T^\star - T_0)^2} \\[1em]
& + \frac{2\,a_2\,\varphi\,A\left(1 + a_2\,(T^\star - T_0)^2\right)^{-1-\phi}}{c^\star} - \frac{4\,a_1^2\,\varphi\,(T^\star - T_0)^2}{\left(1 + a_1\,(T^\star - T_0)^2\right)^2} \\[1em]
& + \frac{4\,a_2^2\,(-1-\phi)\,\phi\,A\left(1 + a_2\,(T^\star - T_0)^2\right)^{-2-\phi}\,(T^\star - T_0)^2}{c^\star}.
\end{aligned}
$$

The eigenvalues of J are calculated according to

$$
e_{1,2,3,4} = \frac{\rho - n}{2} \pm \sqrt{\left(\frac{\rho - n}{2}\right)^2 - \frac{K_1}{2} \pm \sqrt{\left(\frac{K_1}{2}\right)^2 - \det J}},
$$

with K_1 defined as

$$
K_1 = \begin{vmatrix} a_{11} & a_{13} \\ a_{31} & a_{33} \end{vmatrix} + \begin{vmatrix} a_{22} & a_{24} \\ a_{42} & a_{44} \end{vmatrix} + 2\begin{vmatrix} a_{12} & a_{14} \\ a_{32} & a_{34} \end{vmatrix},
$$

with a_{ij} the element of the ith column and jth row of the matrix J (see Dockner and Feichtinger 1991). It is immediately seen that $K_1 < 0$ holds in our model. In Dockner and Feichtinger (1991) it is shown that $K_1 > 0$ is a necessary condition for a Hopf bifurcation, which leads to limit cycles. Consequently, a Hopf bifurcation can be excluded in the social optimum. Furthermore, the eigenvalues are symmetrical around $(\rho - n)/2$ so that

Table 11.5 Steady-State Values and Eigenvalues in the Social Optimum for $a = 1.65\,10^{-3}$, $a_1 = 0.035$, and $\varphi = 0.035$.

a_2	ϕ	$T^\star$	$M^\star$	$E^\star$	Eigenvalues
0.025	0.025	291.1	1.77	0.362	0.005 ± 6.84828, 0.005 ± 0.107673
0.05	0.05	289.3	1.2	0.246	0.005 ± 6.71707, 0.005 ± 0.179849

Table 11.6 Steady-State Values and Eigenvalues in the Social Optimum for $a = 1.65\,10^{-3}$, $m_1 = 0.07$ and $\varphi = 0.07$.

m_2	ϕ	$T^\star$	$M^\star$	$E^\star$	Eigenvalues
0.025	0.025	290.9	1.68	0.344	0.005 ± 6.83093, 0.005 ± 0.110683
0.05	0.05	289.3	1.2	0.244	0.005 ± 6.71475, 0.005 ± 0.182964

there are at most two negative eigenvalues or two eigenvalues with negative real parts. This depends on the signs of det J and of K_1. However, we cannot derive more concrete results for this analytical model.   □

This proposition shows that the BGP in the social optimum is a saddle point, and a Hopf bifurcation generating persistent limit cycles is not possible. However, we cannot answer whether the eigenvalues are real or complex conjugate. Therefore, to gain further insight in the structure of the social optimum, we compute steady-state values[16] and the eigenvalues of the Jacobian matrix at the steady state for the numerical examples in 11.1. The results for the case $a = 1.65\,10^{-3}$ and $a_1 = 0.035$ and $\varphi = 0.035$ are shown in table 11.5.

It can be seen that the higher the damage caused by a temperature increase with respect to aggregate output, the smaller the optimal increase in GHG emissions and in the average global surface temperature. Note that the smaller increase in GHGs is due to higher abatement spending, $B^\star$, relative to gross emissions, $aK^\star$. The same outcome is obtained when the direct damage concerning utility, caused by a higher surface temperature increase, is set higher. This is seen in table 11.6, where we set $m_1 = 0.07$ and $\varphi = 0.07$.

In table 11.7, finally, we look at the optimal steady state values of GHG emissions and of the temperature in an economy with a less polluting technology, that is, with a lower value of a. It is seen that the emissions are lower compared to the situation with a more polluting technology shown in table 11.6, and consequently, the optimal levels of GHG concentrations and of the temperature increase in steady state are lower.

---

[16] The steady-state values were computed using Newton's method.

Table 11.7 Steady-State Values and Eigenvalues in the Social
Optimum for $a = 1.65\,10^{-4}$, $m_1 = 0.035$, and $\varphi = 0.035$.

m_2	ϕ	$T^\star$	$M^\star$	$E^\star$	Eigenvalues
0.025	0.025	288.8	1.08	0.22	$0.005 \pm 6.67681, 0.005 \pm 0.277638$
0.05	0.05	288.5	1.02	0.208	$0.005 \pm 6.64358, 0.005 \pm 0.52092$

The analysis of the social optimum demonstrated that the increase in temperature is smaller the higher the damage caused by the temperature increase. This holds because abatement spending is larger relative to gross emissions for high damages. This result is not too surprising.

Further, our analysis shows that the less polluting the technology, the smaller are emissions after abatement in steady state, and consequently, the smaller the temperature increase. This implies that economies with clean technologies should emit less compared to economies with less clean technologies.

The intuition for this result is as follows. The economy receives utility from consumption and disutility from the temperature increase. The latter acts both directly by affecting utility and indirectly by reducing aggregate production. If the technology in use is relatively clean, it is cheaper to avoid the increase in temperature compared to economies with relatively polluting technologies. Therefore, economies with a less polluting technology have a cost advantage compared to other economies and should have smaller temperature increases.

It should be repeated that the result derived here holds only if deviations of the temperature form the preindustrial level affect the productivity of the economy. In addition, it is only relevant for the social optimum. This implies that this result is not necessarily true for the second-best solution, where the government sets the tax rates taking into account the individual's optimal decisions. If the tax rates are set such that the market economy replicates the social optimum, as stated in proposition 20, this result also holds for the market economy.

11.3 MODELING NONLINEAR FEEDBACK EFFECTS OF THE RISE IN TEMPERATURE

In the last section, we saw that both the competitive economy and the social optimum are characterized by a unique BGB that is saddle point stable. Here we want to study the dynamics of the model when we allow for a nonlinear feedback effect of the temperature increase as in Greiner and Semmler (2005).

In chapter 8 we introduced the basic energy balance model, according to which the change in the average surface temperature on Earth is given by

$$\frac{dT(t)}{dt} c_h \equiv \dot{T}(t) c_h = S_E - H(t) - F_N(t), \; T(0) = T_0. \qquad (11.37)$$

Furthermore, the difference $S_E - H$ can be written as $S_E - H = Q(1 - \alpha_1)\alpha_2/4$, with $Q = 1367.5 \, Wm^{-2}$ the solar constant, $\alpha_1 = 0.3$ the planetary albedo, determining how much of the incoming energy is reflected by the atmosphere, and α_2 ($\alpha_2 = 0.3$) takes into account that a part of the energy is absorbed by the surface of the Earth. Both α_1 and α_2 were assumed to be constant.

However, according to Henderson-Sellers and McGuffie (1987) and Schmitz (1991) the albedo of the Earth is a function that negatively depends on the temperature on Earth. This holds because deviations from the equilibrium average surface temperature have feedback effects that affect the reflection of incoming energy. Examples of such feedback effects are the ice-albedo feedback mechanism and the water vapor "greenhouse" effect (see Henderson-Sellers and McGuffie 1987, chap. 1.4). With higher temperatures a feedback mechanism occurs, with the areas covered by snow and ice likely to be reduced. This implies that a smaller amount of solar radiation is reflected when the temperature rises, tending to increase the temperature on Earth further. Therefore, Henderson-Sellers and McGuffie (1987, chap. 2.4) and Schmitz (1991, p. 194) propose a function as shown in figure 11.1.

Figure 11.1 shows $1 - \alpha_1(T)$, that part of energy that is not reflected by Earth. For the average temperature smaller than T_l, the albedo is a constant, then the albedo declines linearly, so that $1 - \alpha_1(T)$ rises, until the temperature reaches T_u, from which point on, the albedo is constant again. Here, we like to point out that other feedback effects may occur, such as a change in the flux ratio of outgoing to incoming radiative flux, for example. However, we do not take into account these effects because the qualitative result would remain the same.

With these considerations, the EBM can be rewritten as

$$\dot{T}(t) c_h = \frac{1367.5}{4} (1 - \alpha_1(T)) - 0.95 \left(5.67 \, 10^{-8}\right)(21/109) \, T^4,$$

$$T(0) = T_0. \qquad (11.38)$$

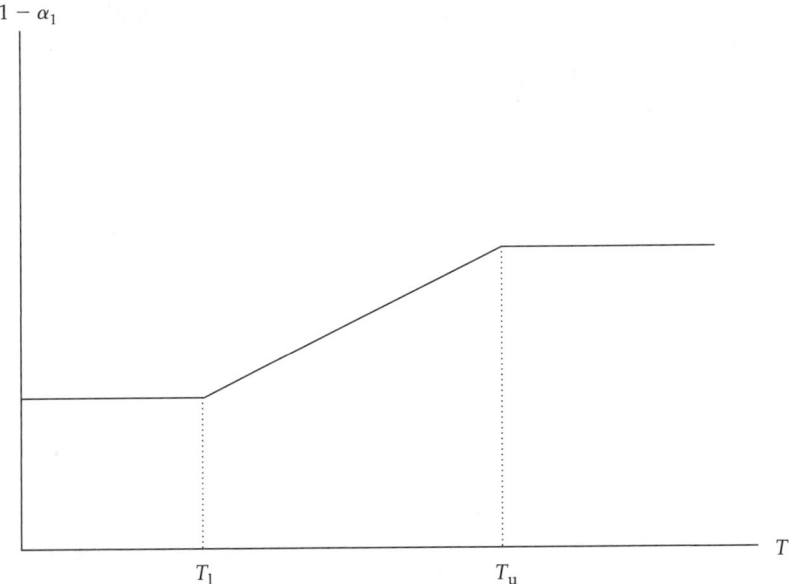

Figure 11.1 Albedo as a Function of the Temperature.

According to Roedel (2001), $(1 - \alpha_1(T)) = 0.21$ holds in equilibrium, for $\dot{T} = 0$, giving a surface temperature of about $288\,K$ which is about $15\,°C$.

Taking into account the emission of GHGs into the atmosphere, the equation can be rewritten as

$$\dot{T}(t)\, c_h = \frac{1367.5}{4}\, (1 - \alpha_1(T)) - 0.95\, \left(5.67\,10^{-8}\right)(21/109)T^4$$

$$+ \beta_1(1 - \xi)\, 6.3\ln \frac{M}{M_o},\, T(0) = T_0. \tag{11.39}$$

11.3.1 The Competitive Economy

In this section we present our economic framework. We start with the description of the structure of our economy, which is basically the same as in the last section.

THE STRUCTURE OF THE ECONOMY

We consider an economy where one homogeneous good is produced. Further, the economy is represented by one individual with household production who maximizes a discounted stream of utility arising from per capita consumption, C, times the number of household members subject to a budget constraint. As to the utility function we assume a logarithmic function $U(C) = \ln C$.[17]

The individual's budget constraint in per capita terms is given by (11.2) and (11.3). Concerning emissions of GHGs, we assume that these are a byproduct of capital used in production and expressed in CO_2 equivalents, so emissions are a function of per capita capital relative to per capita abatement activities. Formally, emissions are described the same equation as in the last section, (11.4).

The agent's optimization problem can be written as

$$\max_{C,B} \int_0^\infty e^{-\rho t} L_0 e^{nt} \ln C \, dt, \tag{11.40}$$

subject to (11.2), (11.3), and (11.4). L_0 is labor supply at time $t = 0$, which we again normalize to unity and grows at constant rate n. Recall that in the competitive economy the agents neither take into account the negative externality of capital, the emission of GHG, nor the positive externalities (i.e., the spillover effects).

To find the optimal solution, we form the current-value Hamiltonian, which is

$$\mathcal{H}(\cdot) = \ln C + \lambda_1((1 - \tau)AK^\alpha \bar{K}^{1-\alpha}D(\cdot) - C - B - \tau_E L^{-1}a^\gamma K^\gamma B^{-\gamma})$$
$$- \lambda_1(\delta + n)K, \tag{11.41}$$

with λ_1 the shadow price of K. Note that we used $E = a^\gamma K^\gamma B^{-\gamma}$.

The necessary optimality conditions are given by

$$\frac{\partial \mathcal{H}(\cdot)}{\partial C} = C^{-1} - \lambda_1 = 0, \tag{11.42}$$

$$\frac{\partial \mathcal{H}(\cdot)}{\partial B} = \tau_E L^{-1} a^\gamma K^\gamma \gamma B^{-\gamma-1} - 1 = 0, \tag{11.43}$$

$$\dot{\lambda}_1 = (\rho + \delta)\lambda_1 - \lambda_1 \left((1 - \tau)A\alpha D(\cdot) - (\tau_E/LK)\gamma a^\gamma K^\gamma B^{-\gamma}\right). \tag{11.44}$$

In (11.44) we state that in equilibrium $K = \bar{K}$ holds. Further, the limiting transversality condition $\lim_{t \to \infty} e^{-(\rho+n)t}\lambda_1 K = 0$ must hold.

Using (11.42) and (11.44) we can derive a differential equation giving the growth rate of per capita consumption. This equation is obtained as

$$\frac{\dot{C}}{C} = -(\rho + \delta) + \alpha(1 - \tau)AD(\cdot) - \gamma \frac{\tau_E}{LK} a^\gamma K^\gamma B^{-\gamma}. \tag{11.45}$$

---

[17] In this section we do not model direct effects of a rise in the temperature on the individual's utility.

Combining (11.43) and (11.4) yields

$$E = \left(\frac{\tau_E}{LK}\right)^{-\gamma/(1+\gamma)} a^{\gamma/(1+\gamma)} \gamma^{-\gamma/(1+\gamma)}. \tag{11.46}$$

Using (11.39) and (8.6), with the numerical parameter values introduced and the equations derived in this section, the competitive economy is completely described by the following differential equations:

$$\dot{T}(t)\, c_h = \frac{1367.5}{4}\, (1 - \alpha_1(T)) - 0.95\, (5.67\, 10^{-8})(21/109)\, T^4$$

$$+ (1 - \xi)\, 6.3 \ln \frac{M}{M_o},\ T(0) = T_0, \tag{11.47}$$

$$\dot{M} = \beta_1 \left(\frac{\tau_E}{LK}\right)^{-\gamma/(1+\gamma)} a^{\gamma/(1+\gamma)} \gamma^{-\gamma/(1+\gamma)} - \mu M,$$

$$M(0) = M_0, \tag{11.48}$$

$$\frac{\dot{C}}{C} = -(\rho + \delta) + \alpha\, (1 - \tau)\, AD(\cdot) - \gamma \left(\frac{\tau_E}{LK}\right)^{1/(1+\gamma)}$$

$$\times a^{\gamma/(1+\gamma)} \gamma^{-\gamma/(1+\gamma)}, \tag{11.49}$$

$$\frac{\dot{K}}{K} = (1 - \tau)\, AD(T - T_0) - \left(\frac{\tau_E}{LK}\right)^{1/(1+\gamma)} a^{\gamma/(1+\gamma)} \gamma^{-\gamma/(1+\gamma)}(1 + \gamma)$$

$$- \frac{C}{K} - (\delta + n),\ K(0) = K_0, \tag{11.50}$$

where $C(0)$ can be chosen by society.

THE DYNAMICS OF THE COMPETITIVE ECONOMY

A balanced growth path or steady state is defined as in the last sections, that is, GHGs and the temperature are constant while all economic variables grow at the same rate.

To study the dynamics of our model, we consider the ratio $c \equiv C/K$, which is constant on a BGP. Thus, our dynamic system is given by the following differential equations

$$\dot{T}(t) = \left(\frac{1367.5}{4}\, (1 - \alpha_1(T)) - 0.95\, (5.67\, 10^{-8})(21/109)\, T^4\right) c_h^{-1}$$

$$+ \left((1 - \xi)\, 6.3 \ln \frac{M}{M_o}\right) c_h^{-1},\ T(0) = T_0, \tag{11.51}$$

$$\dot{M} = \beta_1 \left(\frac{\tau_E}{LK} \right)^{-\gamma/(1+\gamma)} a^{\gamma/(1+\gamma)} \gamma^{-\gamma/(1+\gamma)} - \mu M, \; M(0) = M_0, \quad (11.52)$$

$$\dot{c} = c \left(n - \rho \right) - (1 - \alpha)(1 - \tau) AD(\cdot) + \left(\frac{\tau_E}{LK} \right)^{1/(1+\gamma)}$$

$$\times a^{\gamma/(1+\gamma)} \gamma^{-\gamma/(1+\gamma)} \right) + c^2, \quad (11.53)$$

where $c(0)$ can again be chosen freely by society.

To study the dynamics of our model we resort to numerical simulation. We start with a recapitulation of the parameter values we employ in the numerical analysis.

We consider one time period to comprise one year. The discount rate is set to $\rho = 0.03$, the population growth rate is assumed to be $n = 0.02$, and the depreciation rate of capital is $\delta = 0.075$. The preindustrial level of GHGs is normalized to one (i.e., $M_0 = 1$), and we set $\gamma = 1$. The income tax rate is $\tau = 0.15$, and the capital share is $\alpha = 0.45$. This value seems to be high. However, if capital is considered in a broad sense (meaning that it also comprises human capital), this value is reasonable. A is set to $A = 0.35$, implying that the social gross marginal return to capital is 35 percent for $T = T_0$.

As to τ_E/LK we set $\tau_E/LK = 0.001$, and a is set to $a = 1.65 \cdot 10^{-4}$. τ_E/LK and a determine the ratio of abatement per capital stock, which is given by $B/K = (a\tau_E/LK)^{0.5}$. With these values B/K takes the value $B/K = 4.1 \cdot 10^{-4}$. For example, in Germany the ratio of abatement spending to capital in 2000 was $9.7 \cdot 10^{-4}$ (see Institut der Deutschen Wirtschaft 2003, tables 2.11 and 8.7). Shortly, we will analyze how different values for τ_E/LK affect the dynamics of our model.

Concerning the damage function $D(\cdot)$, we assume function (11.18), which we now denote by $D(\cdot)$, that is,

$$D(\cdot) = \left(m_1 (T - T_0)^2 + 1 \right)^{-\phi}, \quad (11.54)$$

with $m_1 = 0.04$ and $\phi = 0.05$. These values imply that a rise of the surface temperature by 3 (2, 1) °C implies a damage of 1.5 (0.7, 0.2) percent. Recall that the IPCC estimates that a doubling of GHGs, which goes along with an increase of the global average surface temperature between 1.5 °C and 4.5 °C, reduces world GDP by 1.5 to 2 percent (see IPCC 1996, p. 218), so that our choice for the parameters seems justified.

As to the albedo, $\alpha_1(T)$, we use a function as shown in figure 11.1. We approximate the function shown in figure 11.1 by a differentiable function. More concretely, we use the function

$$1 - \alpha_1(T) = k_1 \left(\frac{2}{\Pi} \right) \text{ArcTan} \left(\frac{\Pi (T - 293)}{2} \right) + k_2. \quad (11.55)$$

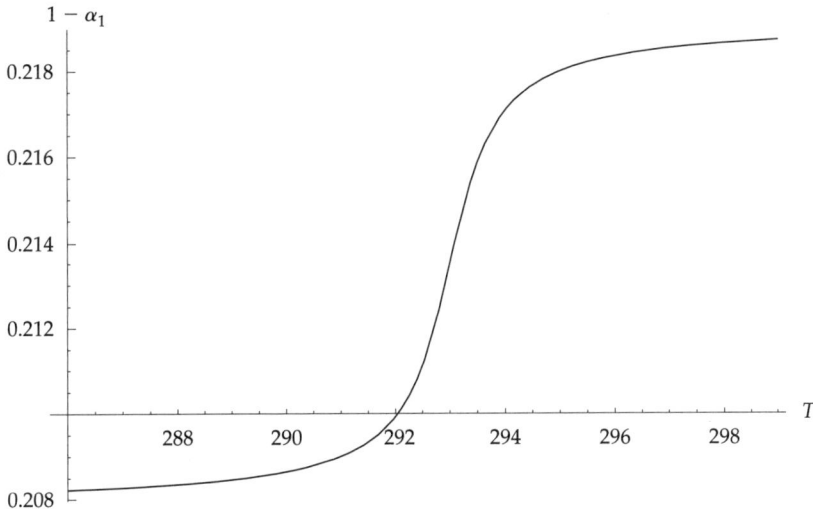

Figure 11.2 Albedo as a Smooth Function of the Temperature.

k_1 and k_2 are parameters that are set to $k_1 = 5.6\,10^{-3}$ and $k_2 = 0.2135$. Figure 11.2 shows the function $(1 - \alpha_1(T))$ for these parameter values.

With (11.55) the preindustrial average global surface temperature is about 287.8 K (for $M = M_o$) and $1 - \alpha_1(\cdot) = 0.2083$. For $T \to \infty$ the expression $1 - \alpha_1(\cdot)$ converges to $1 - \alpha_1(\cdot) = 0.2191$, which corresponds to an increase of about 5 percent.

To get insight into our model, we first note that on a BGP the GHG concentration and the average global surface temperature are completely determined by the emission tax rate τ_E/LK. This holds because this ratio determines optimal abatement spending via (11.43). The global surface temperature on the BGP, then, gives the ratio of consumption to capital and the balanced growth rate, g. Solving (11.52) $= 0$ with respect to M and inserting the result in (11.51) $\equiv dT$ gives a function as shown in figure 11.3.

One realizes that there are three solutions for $dT = 0$. Table 11.8 gives the steady-state values for T^* and c^* and the balanced growth rate, g, as well as the eigenvalues of the Jacobian matrix corresponding to (11.51)–(11.53).

Table 11.8 shows that the first and third long-run steady states (I and III) are saddle point stable, while the second is unstable, with the exception of a one-dimensional stable manifold. Thus, there are two possible long-run steady states to which the economy can converge. The first one implies a temperature increase of about 3.7 °C and a balanced growth rate of about 2.6 percent; the other BGP corresponds to a temperature increase of about 6.2 °C and a balanced growth rate of about

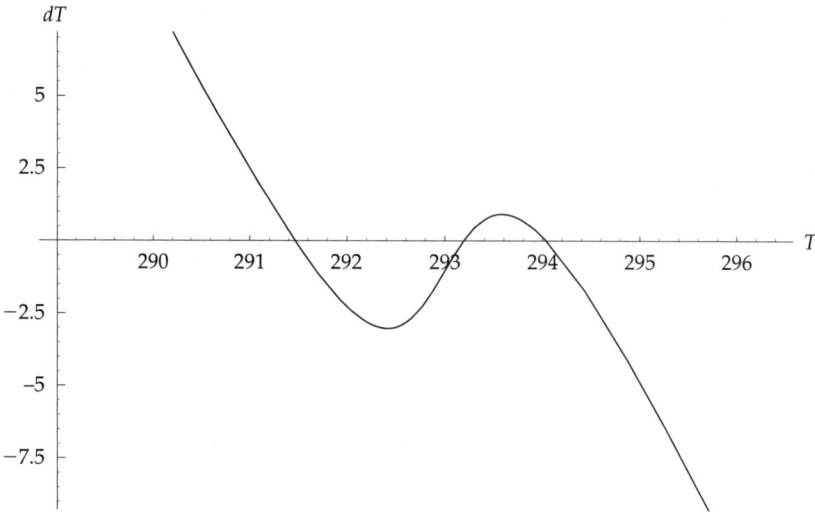

Figure 11.3 Multiple Steady States in the Long Run.

Table 11.8 Steady-State Values, Balanced Growth
Rate, and Eigenvalues for the Competitive
Model with $\tau_E/LK = 0.001$.

Steady State	T^*	c^*	g	Eigenvalues
I	291.5	0.1697	2.6%	$-4.99, 0.17, -0.1$
II	293.2	0.167	2.3%	$4.76, 0.167, -0.1$
III	294	0.1657	2.2%	$-3.55, 0.166, -0.1$

2.2 percent. $1 - \alpha_1(\cdot)$ takes the value 0.2093 for $T^* = 291.5$ and 0.2171
for $T^* = 294$, showing that the quantitative decrease in the albedo does
not have to be large for the occurrence of multiple equilibria. Our result
suggests that there exists a threshold such that the initial conditions
determine whether it is optimal to converge to steady state I or III.

ROBUSTNESS AND COMPARATIVE STATIC RESULTS

The last section demonstrated that there may exist a threshold for the
competitive economy that determines whether it is optimal to converge
to the long-run equilibrium that corresponds to a relatively small rise in
the temperature or to the one with a large temperature increase. Here
we address the question of how robust this result is with respect to
the emission tax ratio τ_E/LK. Furthermore, we want to undertake some
welfare considerations for the economy on the BGP.

We also point out that in the very long run, when fossil fuels will be exhausted, the problem of global warming does not exist any longer. However, our approach models an economy where fossil fuels are an important input factor in the production process. Studying the model along the BGP, then, implies that the economy is successful in stabilizing emissions at a constant but higher level and that the convergence speed is sufficiently high. Of course, the BGP is only reached for $t \to \infty$, but nevertheless, the BGP may be a good approximation if the deviations from it are only small.

Varying the emission tax rate τ_E/LK affects the position of the dT curve in figure 11.3, thus determining the equilibrium temperature and possibly the number of equilibria. A rise in τ_E/LK shifts the dT curve downward and to the left, implying a decrease of the temperature(s) on the BGP. Further, for a sufficiently high value of τ_E/LK, only one equilibrium exists. For example, raising τ_E/LK to $\tau_E/LK = 0.0011$ gives a unique long-run BGP with a steady-state temperature of 291.8 K. This equilibrium is saddle point stable (two negative real eigenvalues). Reducing τ_E/LK to $\tau_E/LK = 0.0008$ also gives a unique BGP with a steady-state equilibrium temperature of 294.8 K. This equilibrium is also saddle point stable (two negative real eigenvalues). This demonstrates that the government choice of the emission tax ratio is crucial in terms of the long-run outcome. This holds for both the temperature in equilibrium and the dynamics of the system.

Presuming the uniqueness of the steady state, we can concentrate on welfare considerations. We limit our investigations to the model on the BGP, although this can only be an approximation. Welfare on the BGP is given by

$$W = \int_0^\infty e^{-(\rho-n)t} \ln(c^\star K^\star e^{gt})dt, \qquad (11.56)$$

which shows that welfare in steady state positively depends on the consumption ratio, $c^\star$; on the balanced growth rate, g, which is determined endogenously; and on $K^\star$, which we normalize to one ($K^\star \equiv 1$). From (11.49) and (11.53) one realizes that τ_E/LK has a negative direct effect on $c^\star$ and on g and a positive indirect effect by reducing the equilibrium surface temperature, which implies smaller damages. This suggests that there exists an inverted U-shaped curve between the emission tax ratio and the growth rate and welfare. To see this more clearly, we calculate the balanced growth rate, $c^\star$, and the average global surface temperature for different values of τ_E/LK and for different damage functions. The results are shown in table 11.9. As for the damage function, we use the parameter values from the last section, $m_1 = 0.04$, $\phi = 0.05$, and, in addition, $m_1 = 0.03$, $\phi = 0.03$. Setting $m_1 = 0.03$ and $\phi = 0.03$ implies that a rise of the surface temperature by 3 (2, 1) °C implies a damage of 0.7 (0.3, 0.09) percent of world GDP.

Table 11.9 Balanced Growth Rate, c^*, and T^* for Different Values of τ_E/LK with $m_1 = 0.04$, $\phi = 0.05$, and $m_1 = 0.03$, $\phi = 0.03$, Respectively.

$m_1 = 0.04, \phi = 0.05$				$m_1 = 0.03, \phi = 0.03$			
τ_E/LK	g	c^*	T^*	τ_E/LK	g	c^*	T^*
0.0011	0.0260	0.1702	291.2	0.0011	0.0273	0.1718	291.2
0.004	0.0280	0.1728	287.8	0.0035	0.0281	0.1729	288.4
0.0055	0.0277	0.1725	287.0	0.0042	0.0280	0.1728	287.8

First, we can see from table 11.9 that the balanced growth rate, g, and the consumption share, c^*, react in the same manner to changes in the emission tax ratio τ_E/LK so that maximizing the balanced growth rate also maximizes welfare. Furthermore, table 11.9 confirms that there exists an inverted U-shaped curve[18] between the emission tax ratio and the balanced growth rate and welfare. For the higher damage ($m_1 = 0.04, \phi = 0.05$) it is optimal to choose the emission tax rate so that the temperature remains at its preindustrial level, implying that the damage is zero. For a lower damage corresponding to the temperature increase ($m_1 = 0.03, \phi = 0.03$) the balanced growth rate is maximized for a value of τ_E/LK that gives an average surface temperature exceeding the preindustrial level. In this case, accepting a deviation from the preindustrial average global surface temperature has positive growth and welfare effects in the long run.

11.3.2 The Social Planner's Problem

In formulating the optimization problem, a social planner takes into account both the positive and negative externalities of capital. Consequently, for the social planner the resource constraint is given by

$$\dot{K} = AKD(T - T_o) - C - B - (\delta + n)K, K(0) = K_0. \tag{11.57}$$

Then the optimization problem is

$$\max_{C,B} \int_0^\infty e^{-\rho t} L_0 e^{nt} \ln C \, dt, \tag{11.58}$$

subject to (11.2), (11.3), (11.4), (11.39), and (8.6), where $D(\cdot)$ is again given by (11.18).

---

[18] We calculated more values that we, however, do not show here.

To find necessary optimality conditions, we formulate the current-value Hamiltonian, which is

$$\mathcal{H}(\cdot) = \ln C + \lambda_2 (A\,KD(T - T_o) - C - B - (\delta + n)K)$$

$$+ \lambda_3 \left(\beta_1\, a^\gamma K^\gamma B^{-\gamma} - \mu M \right) + \lambda_4\, (c_h)^{-1} \cdot \left(\frac{1367.5}{4}(1 - \alpha_1(T)) \right.$$

$$\left. - (5.67 10^{-8})(19.95/109)T^4 + (1 - \xi)6.3 \ln \frac{M}{M_o} \right), \qquad (11.59)$$

with $\alpha_1(T)$ given by (11.55). λ_i, $i = 2, 3, 4$, are the shadow prices of K, M, and T, respectively, and $E = a^\gamma K^\gamma B^{-\gamma}$. Note that λ_2 is positive 2nd λ_3 and λ_4 are negative.

The necessary optimality conditions are obtained as

$$\frac{\partial \mathcal{H}(\cdot)}{\partial C} = C^{-1} - \lambda_2 = 0, \qquad (11.60)$$

$$\frac{\partial \mathcal{H}(\cdot)}{\partial B} = -\lambda_3\, \beta_1\, a^\gamma\, K^\gamma\, \gamma B^{-\gamma - 1} - \lambda_2 = 0, \qquad (11.61)$$

$$\dot{\lambda}_2 = (\rho + \delta)\,\lambda_2 - \lambda_2\, A\, D(\cdot) - \lambda_3\, \beta_1\, \gamma\, a^\gamma\, K^{\gamma - 1}\, B^{-\gamma}, \qquad (11.62)$$

$$\dot{\lambda}_3 = (\rho - n)\,\lambda_3 + \lambda_3\, \mu - \lambda_4\, (1 - \xi)\, 6.3\, c_h^{-1}\, M^{-1}, \qquad (11.63)$$

$$\dot{\lambda}_4 = (\rho - n)\,\lambda_4 - \lambda_2\, A\, K\, D'(\cdot) + \lambda_4\, (c_h)^{-1}\, 341.875\, \alpha_1'(\cdot)$$

$$+ \lambda_4 \left(5.67\, 10^{-8}(19.95/109)\, 4\, T^3 \right) (c_h)^{-1}, \qquad (11.64)$$

with $\alpha_1' = -k_1(1 + 0.25\Pi^2(T - 293)^2)^{-1}$. Furthermore, the limiting transversality condition $\lim_{t \to \infty} e^{-(\rho + n)t}(\lambda_2 K + \lambda_3 T + \lambda_4 M) = 0$ must hold.

Comparing the optimality of the competitive economy with that of the social planner demonstrates how the government has to set taxes to replicate the social optimum. Setting (3.112) = (3.130) shows that τ_E/LK has to be set such that $\tau_E/LK = \beta_1(-\lambda_3)/(\lambda_2 K)$ holds. Further setting the growth rate of per capita consumption in the competitive economy equal to that of the social optimum gives $\tau = 1 - \alpha^{-1}$, which is the same result as in the last section.

From (11.60) and (11.61) we get

$$\frac{B}{K} = \left(c\,(-\lambda_3)\, \beta_1\, \gamma\, a^\gamma \right)^{1/(1+\gamma)}, \qquad (11.65)$$

with $c \equiv C/K$. Using (11.65), (11.60), and (11.62) the social optimum is completely described by the following system of autonomous differential equations

$$\dot{C} = C\left(A D(\cdot) - (\rho + \delta) - ((C/K)(-\lambda_3)\beta_1 \gamma a^\gamma)^{1/(1+\gamma)}\right), \qquad (11.66)$$

$$\dot{K} = K\left(A D(\cdot) - \frac{C}{K} - ((C/K)(-\lambda_3)\beta_1 \gamma a^\gamma)^{1/(1+\gamma)} - (\delta + n)\right),$$

$$K(0) = K_0, \qquad (11.67)$$

$$\dot{M} = (C/K)^{-\gamma/(1+\gamma)}(-\lambda_3)^{-\gamma/(1+\gamma)}\beta_1^{1/(1+\gamma)}\gamma^{-\gamma/(1+\gamma)}a^{\gamma/(1+\gamma)} - \mu M,$$

$$M(0) = M_0, \qquad (11.68)$$

$$\dot{T} = c_h^{-1}\left(341.875(1 - \alpha_1(T)) - 5.67\,10^{-8}(19.95/109)T^4\right.$$

$$\left. +6.3\,(1 - \xi)\ln\frac{M}{M_o}\right), T(0) = T_0, \qquad (11.69)$$

$$\dot{\lambda}_3 = (\rho - n)\lambda_3 + \lambda_3\mu - \lambda_4(1 - \xi)6.3\,c_h^{-1}M^{-1}, \qquad (11.70)$$

$$\dot{\lambda}_4 = (\rho - n)\lambda_4 - A\frac{K}{C}D'(\cdot) + \lambda_4(c_h)^{-1}341.875\,\alpha_1'(\cdot)$$

$$+ \lambda_4\left(5.67\,10^{-8}(19.95/109)c_h^{-1}4\,T^3\right). \qquad (11.71)$$

As for the competitive economy, a BGP is given for variables $T^\star, M^\star, \lambda_3^\star$, $\lambda_4^\star$, and $c^\star$ such that $\dot{T} = \dot{M} = 0$ and $\dot{C}/C = \dot{K}/K$ holds, with $M \geq M_o$. It should be noted that $\dot{T} = \dot{M} = 0$ implies $\dot{\lambda}_3 = \dot{\lambda}_4 = 0$.

To study the dynamics, we proceed as follows. Because $\dot{C}/C = \dot{K}/K$ holds on the BGP, we get from (11.67) and (11.66) $c^\star = \rho - n$. Next, we set $\dot{M} = 0$, giving $M = M(\lambda_3, \cdot)$. Inserting $M = M(\lambda_3, \cdot)$ in $\dot{\lambda}_3$ and setting $\dot{\lambda}_3 = 0$ yields $\lambda_4 = \lambda_4(\lambda_3, \cdot)$. Using $M = M(\lambda_3, \cdot)$ and $\lambda_4 = \lambda_4(\lambda_3, \cdot)$ and setting $\dot{T} = 0$ gives $\lambda_3 = \lambda_3(T, \cdot)$. Finally, inserting $\lambda_3 = \lambda_3(T, \cdot)$ in $\dot{\lambda}_4$ gives a differential equation that only depends on T and a $T^\star$ such that $\dot{\lambda}_4 = 0$ holds and gives a BGP for the social optimum.

For the parameter values employed in the last section with $m_1 = 0.04, \phi = 0.05$ in the damage function shows that there exists a unique BGP that is saddle point stable (two negative real eigenvalues). The temperature and the GHG concentration are $T^\star = 287.9$ and $M^\star = 1.02$, implying a temperature increase of $0.1°C$.

However, this result depends on the damage function. For extremely small damages going along with global warming, we get a different outcome. For example, with $m_1 = 0.004, \phi = 0.004$, a temperature increase of $3°C$ reduces worldwide GDP by merely 0.014 percent. With theses values we get three equilibria where two are saddle point stable and one is unstable. The temperatures on the BGPs are $T_1^\star = 292$, $T_2^\star = 294.3$, and $T_3^\star = 295.4$. The eigenvalues of the Jacobian matrix, $\mu_i, i = 1, 2, 3, 4$, corresponding to (11.68)–(11.71) are $\mu_{11} = 3.37, \mu_{12} = -3.36, \mu_{13} = 0.31$,

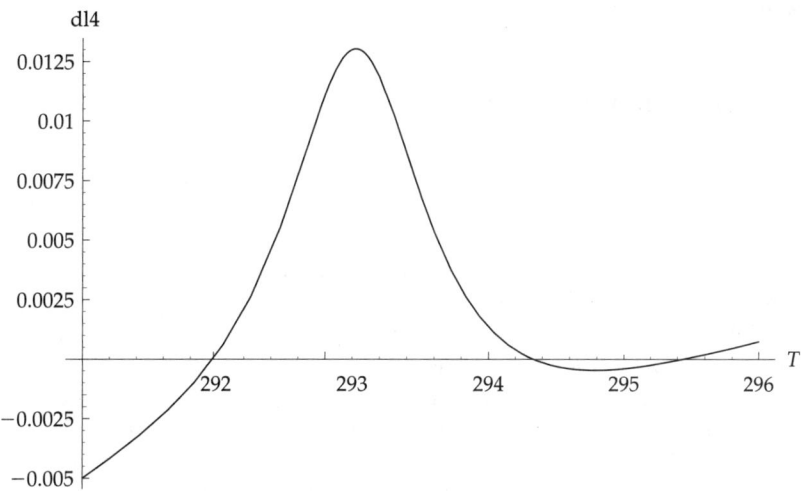

Figure 11.4 Multiple equilibra in the Social Optimum for $m_1 = 0.004$, $\phi = 0.004$.

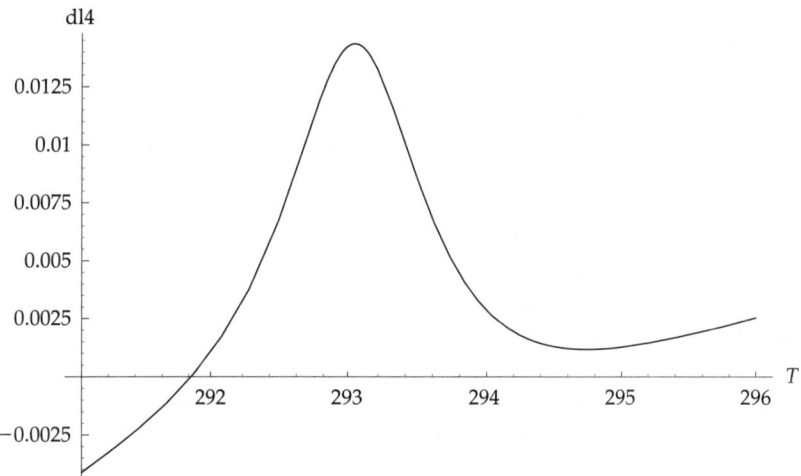

Figure 11.5 Unique Equilibrium in the Social Optimum for $m_1 = 0.004$, $\phi = 0.005$.

$\mu_{14} = -0.3$ for $T = T_1^\star$; $\mu_{12} = 4.7$, $\mu_{22} = -4.69$, $\mu_{23} = 0.005 + 0.12\sqrt{i}$, $\mu_{24} = 0.005 - 0.12\sqrt{i}$ for $T = T_2^\star$; and $\mu_{34} = 6.34$, $\mu_{32} = -6.33$, $\mu_{33} = 0.07$, $\mu_{34} = -0.06$ for $T = T_3^\star$. If the damage of the temperature increase is slightly larger, then the long-run BGP is again unique. Setting $a_1 = 0.004$, $\psi = 0.005$ we get $T^\star = 291.8$, and this equilibrium is saddle point stable. Figures 11.4 and 11.5 illustrate the two situations.

12

Concluding Remarks

In this part we have studied the interrelation between anthropogenic global warming and economic growth assuming a simple descriptive model of endogenous growth. Using simulations, we have seen that increases in abatement spending may yield a win-win situation. That means a rise in abatement activities both reduces greenhouse gas emissions and raises economic growth. This holds for both the balanced growth rate and for the growth rate of GDP on the transition path. Of course, the result that a win-win situation may be observed crucially depends on the damage caused by the temperature increase. The damage function we used was well in line with the damage reported by IPCC studies, so our outcome cannot be dismissed as purely academic.

Furthermore, we have seen that a situation may exist where maximum growth is obtained if the average global surface temperature is reduced to its preindustrial level. This outcome, however, depends on the growth model employed. So in the AK model, maximum growth was obtained for an average global surface temperature that is higher than the preindustrial level.

Assuming a logarithmic utility function, we computed the second-best value for the share of abatement spending, and we have seen that the cleaner the production technology, the smaller the temperature increase should be, compared to economies with more polluting technologies, although in the latter a higher share should be spent for abatement. The same outcome has been obtained for the social optimum. This implies that developing countries countries with less clean technologies spend relatively more for abatement, but nevertheless emit more GHGs than developed countries due to their less clean technology. Further, the abatement share in the social optimum is higher than in the second-best solution, as was expected.

Similar results could also be observed when we allowed for a world with heterogeneous regions concerning the production technology and damages caused by global warming. Our analysis in section 10.3 showed (among other things) that countries with more polluting technologies and higher damages should spend a higher share of GDP for abatement, but nevertheless, may still emit more than countries with cleaner technologies and smaller damages. This holds for both the noncooperative and the cooperative world. Furthermore, the study showed that countries with higher damages from the temperature increase emit less when instruments are set such that marginal damages are equal between regions, compared to the solution of the dynamic optimization problem with cooperation among regions.

In chapter 11 we allowed for optimizing agents and then analyzed both the competitive economy and the social optimum and derived tax rates, which make the competitive economy replicate the social optimum. Using simulations, we could derive the following results.

A situation can again be observed where an increase in the emission tax reduces the temperature increase and raises both economic growth and welfare. Such a win-win situation or double dividend is more likely the higher the damage caused by the increase in temperature concerning aggregate output and the smaller the initial level of the emission tax rate. Note that the emission tax ratio will be smaller with less polluting technology employed. Consequently, economies with a cleaner technology are more likely to experience a double dividend when the emission tax is raised to reduce atmospheric GHG concentrations.

Further, we found that an increase in the emission tax rate may reduce the balanced growth rate but nevertheless lead to higher welfare. This outcome is due to the fact that we allowed for direct disutility effects of an increase in the temperature. This result is more likely the higher the direct negative effect of a temperature increase on utility. Furthermore, for a given damage function, this effect is more likely the smaller the intertemporal elasticity of substitution of consumption.

In the social optimum the increase in temperature is smaller the higher the damage caused by the temperature increase concerning aggregate output and utility, which is not too surprising.

Furthermore, emissions are smaller (and consequently the increase in temperature is lower) the less polluting the technology in use is. The reason for this result is that it is cheaper for economies with cleaner production technologies to avoid the damage caused by the temperature increase. Thus, countries with cleaner technologies should emit less than countries with more polluting technologies. This means that it is optimal for developing countries, which have relatively polluting technologies, to emit more than industrialized countries with relatively clean technologies. This result is quite robust because it was also obtained for all other models in this part.

In section 11.3, finally, we analyzed the dynamics of our endogenous growth model with global warming, taking into account that the albedo of the Earth may depend on the average global surface temperature. Given this assumption, we could demonstrate that the competitive economy can be characterized by multiple long-run BGPs. In this case, the long-run outcome depends on the initial conditions of the economy.

We should like to point out that the change in the albedo does not have to be large to generate this outcome. Our example showed that even a quantitatively small decrease in the albedo may generate multiple equilibria. The existence of the nonlinear feedback effect of a higher temperature influencing the albedo of the Earth leads to this result. Further, other feedback effects, for example, a change in the ratio of emitted to reflected radiative flux, leading to nonlinearities would produce the same qualitative outcome.

The government plays an important role in our models because the choice of the emission tax ratio affects not only the temperature change in equilibrium but also the dynamics of the competitive economy. So the emission tax, and thus the abatement share, is crucial as to whether the long-run BGP is unique or whether there exist several BGPs. The social planner's problem is characterized by a unique BGP for plausible damages going along with global warming. However, if the damages caused by the temperature increase are very small, the social optimum may also generate multiple equilibria and possibly thresholds.

Overall, in terms of government actions, we obtain in the model results similar to those in part I. A zero emission tax is not necessarily welfare-improving. Because there are negative externalities arising from private activities, a government emission tax increases incentives for private abatement activities. As we have demonstrated, by presuming a unique steady state, there exists a growth- and welfare-maximizing, emission tax, and the optimal emission tax is not necessarily zero. However, we want to note that if the presumption of a unique steady state does not hold, bifurcations to multiple equilibria could occur. In this case, however, the problem of an "optimal" tax scheme is more difficult to treat because the welfare effects of tax schemes are more complex.[1]

An interesting and important topic in the analysis of global warming, which we did not take into account, is technical change. Buonanno et al. (2003) and Popp (2003) consider induced technical change in their approaches and demonstrate that this has far-reaching consequences. For example, Popp (2003) shows that integrating technical change in the approach by Nordhaus and Boyer (2000) drastically reduces the cost of abatement and raises welfare gains of lower GHG emissions. However, that model is not an endogenous growth model. As to future research, it would be interesting to extend the model by developing a more elaborate endogenous growth model that incorporates endogenous technical change affecting production and emissions going along with the production process.[2]

---

[1] For a further study of tax rates in the presence of multiple equilibria, see Grüne et al. (2005).

[2] A promising approach working with one representative country is the paper by Gerlagh (2004).

PART III
Depletion of Resources and Economic Growth

13

Introduction and Overview

Now we deal with another topic that has become important with the rise of a global economy. Global economic growth and strong growth in some particular regions of the world economy, for example, the United States, Asia, and some Latin American countries, have given rise to a high demand for nonrenewable and renewable resources. Those resources are in high demand as inputs in the global growth process. Here, too, we can observe a strong externality produced across generations. When current generations deplete resources, future generations will not have those resources at their disposal.

Chapter 14 concentrates on nonrenewable resources and studies models on sustainable economic growth with resource constraints. We also explore to what extent resource constraints can be overcome by substitution and technological change. We examine the problem of intergenerational equity and the different criteria that have been suggested in the literature. Chapter 14 also gives a presentation of stylized facts on exhaustible resources and an estimation of a basic model with resource constraints. We use U.S. time series data. The estimated years left until depletion and the empirical trends of the ratios of capital stock and consumption to resources seem to indicate that there is likely to be a threat to sustainable growth in the future. In our estimation, we obtain parameter values that help interpret the extent to which growth with exhaustible resources is sustainable or nonsustainable. Yet because the time series data are rather short and there are great uncertainties concerning the size of existing stocks, chapter 14 should be seen more as methodological device to estimate resource models, rather then giving definite answers.

Chapter 15 turns to renewable resources and studies the interaction of economic agents, extracting renewable resources, and the resource dynamics as well as the fate of the resources in the long run. We thus present only theoretical work. We demonstrate—partly analytically and partly numerically—that traditional results in resource economics on the depletion of resources obtained from the study of only one resource do not carry over to ecologically interacting resources. As in the traditional approach, and in the approaches of the previous two parts, we employ dynamic optimization. We use a new analytical method. The limiting behavior of the resources is first studied analytically by letting the discount rate approach infinity. A numerical study is then undertaken by means of a dynamic programming algorithm to explore the fate of the resources for various finite discount rates. The relation of our results to results in optimal growth theory is also discussed. We show that if resources are considered in their dynamic interaction, when extracted by economic agents, new insight can be gained that were not feasible in the traditional way of studying resource exploitation (when only one resource was considered). We also study the fate of renewable resources when they interact locally or globally and explore how economic competition may accelerate the depletion of resources. This method also allows us to point out an approach to pricing of environmental assets.

Finally, in chapter 16, we study the problem of how and what type of regulation could be implemented if resources are threatened to be exhausted. This chapter presents a prototype intertemporal model of resource extraction with three state variables and one control variable. The three state variables are the stock of the resource, the capacity of the extractive industry, and its evolution of debt. As an example, for a regulatory control variable we study the effect of a regulatory tax rate. As shown for certain growth functions of the extracted resource, the optimal tax rate is cyclical. We then conjecture that the results may carry over to other regulatory policies.

14

Nonrenewable Resources

14.1 INTRODUCTION

After World War II, many OECD countries exhibited strong economic growth. Since the early 1970s there was growing concern that economic growth has increasingly depleted the available resources.[1] As economic decisions are restricted by the finiteness of natural resources attention has been devoted to the question whether it is possible and desirable to continue present patterns of economic growth.

Economists such as Meadows et al. (1972) or Daly (1987) have put forward pessimistic predictions about a "sudden and uncontrollable decline in both population and industrial capacity" if no "conditions for ecological and economic stability that is sustainable far into the future" are established.[2] This concern about sustainable growth was soon echoed by others. Yet other economists, like Beckerman (1974), held a more optimistic view, stating that technological progress and the discovery of new substitutes make continued economic growth possible. A general consensus of the economic growth debate is that there are trade-offs among environmental and economic goals. The agreement is that economic activity that ignores the biological or social system is not sustainable. There exist many different definitions of sustainability, and all of them have two points in common. First, they recognize that resource and environmental constraints affect the patterns of development and consumption in the long run. Second, they are concerned about equity between generations (intergenerational equity). One of the most famous definitions is stated by the Brundtland Commission (1987): "Sustainable development is development that meets the needs of the present without compromising the ability of future generations to meet their own needs." Similarly Solow (1974) defines sustainablity as "an obligation to conduct ourselves so that we leave to future the option or the capacity to be as well off as we are." Pearce et al. (1990) point out that "natural capital stock should not decrease over time," whereas Pezzey (1961) defines sustainable economic growth as "non-declining output or consumption over time" and sustainable economic development as "non-declining utility over time."

---

[1] This chapter is based on joint work with Almuth Scholl; see Scholl and Semmler (2002).
[2] Meadows et al. (1972), p. 23.

As to the sustainability of natural resources, one can distinguish between renewable and nonrenewable resources. Dynamic models on renewable resources can be found in Clark (1990), Sieveking and Semmler (1997), Semmler and Sieveking (1994b), and in chapter 15 of this book. In those works, theorems on the sustainability of resource economics that have been developed by studying the fate of renewable resources when resources interact as an ecological system. Furthermore, in Semmler and Sieveking (2000), credit-financed extractions of resources are considered and the fate of resources studied.

The current chapter deals with dynamic models with exhaustible resources. We discuss protype growth models that incorporate and study the consequences of finitely available exhaustible resources. Some of the problems studied here will also arise in the case of renewable resources, for example, the problem of intergenerational justice. The remainder of this chapter is organized as follows. In section 14.2 we survey growth models with natural resource constraints. Section 14.3 discusses the problem of intergenerational justice. Sections 14.4 and 14.5 present stylized facts of exhaustible resources. Section 14.6 presents the estimation of our growth model. An appendix concludes this chapter.

14.2 ECONOMIC GROWTH WITH RESOURCE CONSTRAINTS

We review growth models that take into account the presence of exhaustible natural resources. The main interest of the analysis is the question of to what extent the economic growth process is restricted by the finiteness of resource stocks and whether sustained consumption and utility levels are feasible.

We first consider a basic model where the exhaustible natural resource is used as an input for the production of a good that is then either consumed or added to the capital stock to enhance future production. Then the consequences of different extensions concerning technological progress are analyzed; finally, it is assumed that natural resources may provide services in preserved states.

Before turning to the description of the models, it is important to have a clear distinction between renewable and exhaustible resources. The main feature of an exhaustible resource is that its growth rate is nil, and it is unrecyclable. Furthermore, it is used up when used as an input in production. To have a meaningful problem, the natural resource must be essential, that is, production is impossible without it.[3]

14.2.1 A Basic Model

Economists like Dasgupta and Heal (1974), Stiglitz (1974), and Solow (1973) analyze the optimal depletion of exhaustible natural resources in

---

[3] Dasgupta and Heal (1979).

the context of a growth model where the resource is used as an input for the production of a composite commodity. The production function F depends on the flow of the exhaustible resource at date t and on the stock of a reproducible good at date t. To obtain greatest possible social welfare, the present value of utility U derived from consumption C_t of the produced good is maximized subject to the evolution of the reproducible capital stock K_t and the constraints imposed by the finiteness of the resource stock S_t:

$$Max \int_0^\infty U(C_t)e^{-\delta t}\, dt \qquad (14.1)$$

s.t.

$$\dot{K}_t = F(K_t, R_t) - C_t$$

$$S_t = S_0 - \int_0^\infty R_t\, dt$$

$$\dot{S}_t = -R_t.$$

δ denotes the discount rate, and R_t is the flow of the exhaustible resource. The initial capital stocks K_0 and S_0 are strictly positive and given. The production function $F(K_t, R_t)$ is assumed to be increasing, strictly concave, twice continuously differentiable, and homogenous of degree unity. The utility function $U(C_t)$ is supposed to be strictly concave, and for $C_t \to 0$ its first derivative is infinity. Here, the extraction of the resource is assumed to be costless.

Solving the maximization problem and combining the optimality conditions yields the following results. For details see appendix A.1. First, along an optimal path the rate of consumption depends on the discount rate δ, on the elasticity of marginal utility of consumption η, and on the marginal productivity of reproducible capital F_K:

$$\frac{\dot{C}_t}{C_t} = \frac{F_K - \delta}{\eta} \qquad (14.2)$$

with $\eta(C_t) = -(C_t U''(C_t)/U'(C_t))$ and $F_K = \partial F(K_t, R_t)/\partial K_t$. Equation (14.2) states that the higher the discount rate, the more the rate of consumption falls over time along an optimal path. Second, along an optimal path the rates of return of exhaustible and reproducible capital are equal:

$$F_K = \frac{\partial F_R}{\partial t} \frac{1}{F_R}, \qquad (14.3)$$

with $F_R = \partial F(K_t, R_t)/\partial R_t$. If the production function is homogenous of degree one, it is possible to set $x_t = K_t/R_t$ with $f(x_t) = F(K_t/R_t, 1)$.

Substituting $F_R = f(x_t) - x_t f'(x_t)$ and $F_K = f'(x_t)$ in equation (3) yields the following capital-resource ratio along an optimal path:

$$\frac{\dot{x}_t}{x_t} = \sigma \frac{f(x_t)}{x_t}, \qquad (14.4)$$

with

$$\sigma = \frac{-f'(x_t)(f(x_t) - x_t f'(x_t))}{x_t f(x_t) f''(x_t)}$$

as the elasticity of substitution between reproducible capital and the exhaustible resource. Equation (14.4) represents the rate at which reproducible capital is substituted for the exhaustible resource. It depends on the elasticity of substitution and the average product per unit of fixed capital.

To conclude whether a positive level of consumption is sustainable over time, Dasgupta and Heal (1974) analyze an economy where output is produced by a CES-production function, that is, the case of a constant elasticity of substitution. There are three cases to mention.

1. $\sigma = 1$ (i.e., the Cobb-Douglas-production function). The exhaustible resource is essential and infinitely valuable at the margin, whereas the asymptotic value of marginal productivity of capital is zero. Solow (1973) concludes that sustained per capita consumption is feasible if the share of capital exceeds that of natural resources.
2. $0 \leq \sigma < 1$. The exhaustible resource is essential but finitely valuable at the margin. Thus, a positive and nondecreasing level of consumption over an infinite time horizon is not sustainable.
3. $\infty > \sigma > 1$. Sustained consumption is feasible because in this case the exhaustible resource is inessential.

14.2.2 Technological Change

The basic model can be augmented by introducing technical change, which makes it easier to find new substitutes to render an essential natural resource inessential. Dasgupta and Heal (1974) assume that technical progress is uncertain: the exact date of discovering a substitute and its detailed characteristics and usefulness are unknown. The new technique is supposed to occur at an unknown date T, which is a random number with an exogenously given probability density function ω_t:

$$Probability \ (T = t) = \omega_t$$

$$\int_0^\infty \omega_t \, dt = 1$$

$$\omega_t > 0$$

To express the situation of uncertainty, the objective is to maximize the expected present value of utility. After some manipulation one obtains the following maximization problem:[4]

$$Max \int_0^\infty [U(C_t)\Omega_t + \omega_t W(K_t, S_t)]e^{-\delta t}\, dt \qquad (14.5)$$

s.t.

$$\dot{K}_t = F(K_t, R_t) - C_t$$

$$S_t = S_0 - \int_0^\infty R_t\, dt$$

$$\dot{S}_t = -R_t,$$

with $\Omega_t = \int_t^\infty \omega_t\, dt$ and $W(K_t, S_t) = Max \int_T^\infty U(C_t)e^{-\delta[t-T]}\, dt$. K_t, C_t, R_t, and S_t are all non-negative, and the initial values K_0 and S_0 are given. Solving the maximization problem, combining the first order conditions, and arguing that at the discovery date of the substitute the then-existing stocks of reproducible and natural capital have no economic value anymore because the new technology is more efficient (i.e., $W_K = W_S = 0$), allows for the following conclusions (for details, see appendix A.2). First, along an optimal path the rate of consumption depends on a modified discount rate:

$$\frac{\dot{C}_t}{C_t} = \frac{F_K - (\delta + \psi_t)}{\eta}, \qquad (14.6)$$

with $\psi_t = \omega_t/\Omega_t$ as the conditional probability of the technological breakthrough at date t given the substitute has not been discovered earlier. The discount rate is modified by the addition of the factor ψ_t, showing the probability of the essential resource becoming inessential as a result of technical progress. Thus, in a situation of uncertainty, the discount rate is higher than in a situation of certainty.[5] Obviously, the equation describing the ratio of capital-resource input is the same as before:

$$\frac{\dot{x}_t}{x_t} = \sigma\frac{f(x_t)}{x_t}. \qquad (14.7)$$

Stiglitz (1974) examines an economy where output is produced by a Cobb-Douglas-production function. He concludes that sustained per capita consumption is feasible, if there is a resource augmenting technical change at any positive rate (for $\sigma \geq 1$). Toman et al. (1993) point out that for the case $\sigma < 1$ sustained per capita consumption is possible if technological progress is high enough.

---

[4] Dasgupta and Heal (1974), p. 20.
[5] See also Sieveking and Semmler (1997).

14.2.3 The Role of Backstop Technology

So far it is assumed that the natural resource is exhaustible, that is, once it is used up, it is impossible to find more, and the extraction is cost-less. As an extension it is now supposed that the resource is available in unlimited quantities but at various grades and various costs. For exam-ple, the ores of a number of metals can be extracted from the deposits currently used that are exhaustible. If they are used up, the metals them-selves can be extracted from the sea or from rock formations, which is much more expensive. Thus, at higher prices the natural resource may be of unlimited availability. Heal (1993) calls this a *backstop technology*. We can incorporate it into the basic model described in section 2.1. The total amount of the resource used at date t is denoted as follows:

$$z(t) = \int_0^\infty R_t \, dt.$$

It is assumed that at date T the conventional deposits are exhausted and a backstop technology takes over. Up to the level z_T the extrac-tion costs rise with cumulative extraction, then the backstop technology is available at a constant cost per unit b. $g(z_t)$ denotes the extraction costs per unit at date t with $\partial g / \partial z_t = g'(z_t) > 0$ for $0 \leq z_t \leq z_T$ and $g(z_T) = b > 0$ for $z_t \geq z_T$.

The maximization problem is solved in two steps (see appendix A.3). First, the situation is examined before current deposits are exhausted (maximization problem [14.8]); second, the situation is examined after the backstop technology has taken over (maximization problem [14.9]).

$$Max \int_0^\infty U(C_t)e^{-\delta t} \, dt \tag{14.8}$$

s.t.

$$\dot{K_t} = F(K_t, R_t) - C_t - g(z_t)R_t$$

$$S_t = S_0 - \int_0^\infty R_t \, dt$$

$$\dot{S_t} = -R_t,$$

where $g(z_t)R_t$ represents the total extraction costs.

$$Max \int_0^\infty U(C_t)e^{-\delta t} \, dt \tag{14.9}$$

s.t.

$$\dot{K_t} = F(K_t, R_t) - C_t - bR_t.$$

The initial capital stocks K_0 and S_0 are strictly positive and given. Computing the conditions along an optimal path of problem (8) yields:

$$\frac{\dot{C}_t}{C_t} = \frac{F_K - \delta}{\eta} \tag{14.10}$$

and

$$F_K = \frac{\partial F_R}{\partial t}\frac{1}{F_R} + \frac{F_K g(z_t)}{F_R}. \tag{14.11}$$

Substituting $F_K = f'(x_t)$ and $F_R = f(x_t) - x_t f'(x_t)$ with $f(x_t) = F(K_t/R_t, 1)$ results in the following capital-resource ratio along an optimal path:

$$\frac{\dot{x}_t}{x_t} = \frac{\sigma f(x_t)}{x_t} + \frac{f'(x_t)}{x_t f''(x_t)}\frac{g(z_t)}{x_t}. \tag{14.12}$$

Condition (14.12) is a generalization of condition (14.4). Here, the equation is augmented by the term $[f'(x_t)/x_t f''(x_t) \times g(z_t)/x_t]$, which reflects the cumulative costs of extraction.

Following Heal (1993), we can draw the following conclusions. During the initial period the lower-cost stocks of the natural resource are exhausted, and the path of the economy is described by problem (14.8) and conditions (14.10) and (14.11). The difference between prices and extraction costs, that is, the user costs, decline until they reach zero at date T, when the backstop technology takes over because the lower-cost stocks are totally used up. From then on, the economy behaves according to problem (14.9), thus, the extraction costs of the natural resource always equal its price.

14.3 THE IMPORTANCE OF INTERGENERATIONAL EQUITY

As apparent from the foregoing models, the depletion of resources generates externalities for future generations. Therefore, the problem of intergenerational equity arises. To study this problem, we first show how natural resources may affect the welfare of the society.

14.3.1 The Amenity Value of Natural Resources

There are two ways a natural resource contributes to society's welfare. The models described so far refer to the first way: the resource is utilized as an input factor for the production of a composite commodity. The second way to serve the well-being of the society is that it may provide valuable services in preserved states, that is scientific, recreational, and aesthetic values. To take into account these so-called amenity values of natural resources, the resource stock S_t is included in the utility function

(Krautkraemer [1985]). The objective is to maximize present value of utility (see appendix A.4):

$$Max \int_0^\infty U(C_t, S_t)e^{-\delta t} \, dt \qquad (14.13)$$

s.t.

$$\dot{K}_t = F(K_t, R_t) - C_t$$

$$S_t = S_0 - \int_0^\infty R_t \, dt$$

$$\dot{S}_t = -R_t,$$

where C_t, K_t, R_t, S_t are all non-negative. $U(C_t, S_t)$ is assumed to be twice continuously differentiable with $U_C = \partial U(C_t, S_t)/\partial C_t > 0$, and $U_S = \partial U(C_t, S_t)/\partial S_t > 0$, $U_{CC} = \partial^2 U(C_t, S_t)/\partial C_t^2 < 0$ and $U_{SS} = \partial^2 U(C_t, S_t)/\partial S_t <$, and $\lim_{C_t \to 0} U_C = \infty$. The production function $F(K_t, R_t)$ has the same properties as before. Solving the maximization problem and combining the first order conditions yield the following conditions for the rate of consumption and for the capital-resource ratio along an optimal path:

$$\frac{\dot{C}_t}{C_t} = \frac{F_K - \delta - (U_{CS}/U_C)R_t}{\eta}, \qquad (14.14)$$

$$\frac{\dot{x}_t}{x_t} = \sigma \frac{f(x_t)}{x_t} + \frac{U_S/U_C}{x_t^2 f''(x_t)}. \qquad (14.15)$$

The amenity services of a natural resource affect the extraction and consumption mainly through contribution to the rate of return to the resource stock. Resource amenities enhance the value of the resource stock. Therefore, the initial price of the resource must be higher than in the previous models. Furthermore, because $f''(x_t) < 0$, the amenity services lower the rate of change in the capital-resource ratio by the term $(U_S/U_C)/x_t^2 f''(x_t)$. The higher the marginal rate of substitution between amenities and consumption, the higher the reduction of the rate of change of the input ratio.

In this framework, where natural environments are valued in their own rights, the so-called Green Golden Rule[6] can be introduced, which is motivated by the Golden Rule of Economic Growth.[7] The Golden Rule of Economic Growth gives the growth path with the highest indefinitely maintainable level of consumption, whereas the Green Golden

[6] Chichilnisky (1996) and Beltratti et al. (1994).
[7] Phelps (1961).

Rule focuses on the highest indefinitely maintainable level of instantaneous utility. Thus, the Green Golden Rule incorporates the aim of sustainability. Formally the rule can be written as:[8]

$$max_{feasible\ paths}\ lim_{t \to \infty} U(C_t, S_t).$$

If the resource is used as an input factor for the production of a composite commodity, in the long run the only constant level of resource input is zero. Because the resource is essential, no output can be produced and consumption is zero. Hence, the highest indefinitely maintainable utility level is feasible if the total initial stock is conserved.

14.3.2 Resources and Intergenerational Equity

Standard growth models that incorporate the concept of sustainability focus on the consequences of natural resource constraints on the long-run pattern of economic development and consumption. To determine intertemporal welfare, recent growth theory has used the concept of discounted utilitarianism, that is, the future is discounted in comparison with the present. Ramsey (1928) states that discounting is "ethically indefensible and arises merely from the weakness of the imagination" because a positive discount rate results in an asymmetric treatment of present and future generations.[9] As for the future damages, the role of the size of the discount rate has become a controversial issue in the recently published Stern report.[10] Many economists took issue with the low discount rate used in the Stern report in the computation of the current cost of future damages arising from climate change.

Discounted future utility from resources, using a high discount rate, neglects intergenerational equity as the second important point of the concept of sustainability. In the following, we give a brief review of some alternative concepts that try to meet the objective of a fair treatment of different generations.

A simple way to account for intergenerational equity is to assume the case of a zero utility discount rate, that is, present and future generations are given the same weight. Another alternative is to apply the "overtaking criterion," as proposed by Weizsäcker (1967), which states that one consumption path is better than another if from some date on total utility of that path is greater. Formally, if

$$\int_0^T U(C_t^1)\, dt \geq \int_0^T U(C_t^2)\, dt.$$

[8] Heal (1998), p. 43.
[9] The discount rate is also important to assess the impact of future damages arising from external effects.
[10] See Stern (2006, 2007).

But applying these approaches gives rise to technical problems. For a zero discount rate the set of attainable values of the integral may be open, and the way of ranking consumption paths according to the overtaking criterion is incomplete.

According to the Rawlsian criterion,[11] intergenerational equity means: maximize the welfare of the less advantaged generation. Formally,

$$max_{feasible\ paths}\ min_{generations\ t}\ (Welfare_t).$$

The consequence of this decision rule is that the welfare level should be the same for all generations. If a later generation enjoys higher welfare, an earlier generation should increase its own welfare at the expense of the later generation and vice versa. Solow (1973) points out that in comparison with the discounted solution based on utilitarianism the Rawlsian criterion will use up the natural resource stock faster. Because the utilitarian rule demands higher savings, earlier generations will have a lower standard of living than the constant max-min rule would generate. The Rawlsian criterion has two main difficulties. First, a society needs an initial capital stock high enough to make a decent standard of living possible, but the explanation of its existence is missing. Second, the rule does not yield a reasonable result if ongoing technical progress is assumed.[12]

More recently, Chichilnisky (1996) defines two axioms for sustainability that deal with the problem of intergenerational equity. The first axiom states that the present generation should not dictate the outcome in disregard for the future. The second axiom states that the long-run future should not dictate the present. Welfare criteria that do satisfy the two axioms are called sustainable preferences. To formulate a criterion that belongs to the class of sustainable preferences, positive weight is placed on the present and on the very long-run properties of a growth path. Formally,

$$\alpha \int_0^\infty U(C_t, S_t)\Delta(t)\ dt + (1 - \alpha)lim_{t\to\infty}(C_t, S_t),$$

where $\alpha \in (0, 1)$. $\Delta(t)$ is any measure with $\int_0^\infty \Delta(t)\ dt = 1$. If $\Delta(t) = e^{-\delta t}$, the first term is just the discounted integrals of utilities. The second term reflects the limiting properties of the utility stream, and it has already been mentioned as the Green Golden Rule solution. The Chichilnisky criterion places more weight on the future than the standard approach

---

[11] Rawls (1972).
[12] A more elaborate version of the Rawls criterion is proposed in Semmler and Sieveking (2000).

of discounting utility but less than the Green Golden Rule. It is possible to apply the Chichilnisky criterion to neoclassical growth models at the aforementioned type,[13] but finding the solution is quite complicated. We therefore leave aside detailed discussions.

This very short review of different welfare criteria has shown that it is very difficult to find approaches that do meet the objective of permitting intergenerational equity and are technically operable at the same time. For that reason, discounted utility is still dominant, as it the technically most convincing approach, though it favors the present over the future.

14.4 ASSESSMENT OF THE FINITENESS OF RESOURCES

Next we discuss some stylized facts that may become relevant in terms of the previously stated theories. Economic theory states that substitution possibilities, technological progress, and the value of the resource in preserved states may prevent the total depletion of natural capital. In this section, the patterns of some selected nonrenewable resources of the U.S. economy are analyzed from 1960 to 1995, which is an interesting time period because due to large oil shocks in the 1970s, big changes took place. We are using these limited time series data to illustrate the methodology that can be used for the estimation of the aforementioned theoretical model. The empirical illustrations serve as background for our estimations in section 14.6. First extraction rates, available resource quantities today and in the future, as well as the time to exhaustion, are examined. We also pursue the question whether there are reasons to argue that improved technology and the development of reproducible substitutes make a sustainable economic development possible.

Two different kinds of natural resources fulfill the main characteristics of exhaustibility: fuel minerals, such as energy resources, and nonfuel minerals, such as metals and industrial minerals. A detailed discussion of the data sources and some rough trends in the time series data can be found in appendix 6. When talking about the limited availability of natural resources, it is important to have clear definitions of the different components of which the total resource stock consists. Figure 14.A1 in appendix A.6 explains the different parts. The reserves of the discovered resources consist of proved reserves and other reserves, such as inferred reserves (field growth), measured reserves, and indicated reserves.[14] Proved reserves are those amounts of the resource that geological and engineering data demonstrate with reasonable certainty to

---

[13] For a detailed analysis see Heal (1995).
[14] Thus, as literature on exhaustible resources has pointed out, there is great uncertainty as to the size of the available stocks; see Nyarko and Olson (1996).

be recoverable in the future from known reservoirs under existing economic and technological conditions. The other reserves consist on one hand of that part of the identified economically recoverable resource that will be added to proved reserves in the future through extensions, revisions, and the discovery of new fields in already discovered regions, and on the other hand on those quantities of the resource that may become economically recoverable in the future from existing production reservoirs through the application of currently available but as yet uninstalled recovery technology.[15]

As noted before, fuel minerals are energy sources such as crude petroleum, coal, and natural gas. To take the time period from the 1960s to the 1990s while the U.S. economy experienced continued economic growth, that is, a rising level of real GDP, total energy consumption has increased by roughly 35 percent (but where also some major oil shocks have occurred). To satisfy increasing energy demand, the production of especially coal and natural gas have risen. To draw conclusions whether the resources are used more efficiently over time methodologically, one can analyze the patterns of production rates per dollar of real GDP, see figures 14.A3–A4 in appendix A.6. Tables 14.A1–A2 in appendix A.6 summarize the results for a 35-year time period.

As exemplified for the time period 1960–95 in table 14.A1, for every energy resource the production rates per GDP have been falling, that is, crude petroleum, coal, and natural gas have been used more efficiently over the 35 years, which may be the result of improved technology. Nonfuel minerals are on one hand metals and on the other hand industrial minerals. Here, only some selected metals (such as copper, iron ore, lead, and zinc) are analyzed. The production rates of these metals behave quite unregulary from 1960 to 1995, and no trends can be determined. Figure 14.A3 and table 14.A2 show that the production rates per dollar real GDP are decreasing.

Note that exhaustible energy resources can be substituted by renewable energy sources, for example, wind and solar power. Figure 14.A4 shows the shares of the different energy sources. Recently, nuclear power and renewable energy together are just 20 percent of total energy production. But the trend is that the production of crude petroleum, coal, and natural gas is decreasing while nuclear and renewable energy production has been increasing over time. Copper, lead, and zinc are metals that can be substituted by aluminium and plastic. For iron ore, there does not exist any substitutes, but as it will be shown later, the reserves of iron ore are very high and will last for the next few centuries.

To draw conclusions about future availability of exhaustible resources, it is necessary to have a closer look at the amount of proved

---

[15] See Energy Information Administration, U.S. Geological Survey.

and other reserves of a resource in comparison with its production rates.

For the exhaustible energy resources crude petroleum and natural gas, we plot the reserve-production ratio for the observed time period, see figure 14.A5. The smaller the ratio, the scarcer the natural resource. Already since the 1960s for petroleum and natural gas the trend is declining, thus, proved reserves will tend to be depleted if, for example, extensions/discoveries of new fields in already discovered regions or new discoveries make reserve additions only minor. This issue will discussed more in detail in section 14.5.

Every year the Energy Information Administration and the U.S. Geological Survey estimate quantities of technically recoverable resource amounts that could be added to the the already proved reserves of the United States. Given the trends over the time period 1960–95, it is interesting to ask how many years it will take to exhaust the current estimated technical recoverable resources quantities. Because the production rates of the energy resources coal and natural gas are steadily increasing during the observed time period, it is assumed that they continue to increase with an average production growth rate. The other resources do not show any clear trend in their production rates, and therefore, it is supposed that production will continue to follow a stable pattern during the years ahead. Tables 14.A3 to 14.A5 summarize the results for the United States. Here again we are less interested in the accuracy of the predicted results but in some rough sketch of some trends.

Although the results of the empirical analysis are restricted by the length of the examined time period and there should be a clear concern with the finiteness of resources. On the other hand, the data may also give reasons not to be too pessimistic because there are other worldwide resources, and in addition, technological progress has made a further development of renewable substitutes for exhaustible resources possible. As the U.S. economy has experienced continued economic growth during the second half of this century, in particular since the 1990s, the production rates of exhaustible resources have risen, and for all analyzed minerals the production per dollar real GDP has declined. Increased efficiency from the consequence of advanced technology and the use of close substitutes are likely explanations for that fact.

Assuming no changes in production trends yields predictions of the number of years left until present estimated resource reserves are depleted. For the United States the estimated reserves of coal and natural gas will last for the next two or three centuries. On the other hand, the present estimated reserves of crude petroleum should raise more concerns. The petroleum resources in the United States will be depleted in decades, for example in 30 to 40 years. Yet there are other worldwide petroleum resources that we discuss next.

14.5 ASSESSMENT OF PETROLEUM RESERVES

As to the share in total world petroleum supply, the Middle East is the crucial supplier. To see just how important it is, consider the following data taken from the Energy Information Administration (January 2003). The data are summarized in table 14.A6 in appendix 6. Total net imports of oil worldwide in 2002–2003 were 24 million barrels per day. The United States accounts for roughly 40 percent of this figure. The OECD Europe has also large import shares, most of it from the Middle East. In terms of percentages, see the figures in brackets in the table 14.A6; the U.S. import share from the Middle East is lower, 20.2 percent as opposed to 35.9 percent in OECD Europe. As can be observed, for the United States, the largest fraction of imports comes from Saudi Arabia. Saudi Arabia is in fact the largest source of oil for all the major OECD countries and also for OECD Europe. For the United States, however, a considerable fraction of imports comes also from Venezuela, which exports nothing to OECD Europe.

This shows that the Middle East is currently important as petroleum supplier. Other important figures are the annual production, in relation to the remaining reserves, and the time to exhaustion. Table 14.A7 summarizes world petroleum reserves, the ratio of production to reserves and time in exhaustion (in brackets). These figures are also from the Energy Information Administration (January 2003).

Overall, we can see that roughly two thirds of the world's proven reserves are in the Middle East. To see how unbalanced this really is note that 25.7 percent of world reserves (and roughly 40 percent of Middle East reserves) are in Saudi Arabia alone. Iraq, second in line, has a share in world oil reserves of about 11 percent.

The first row in table 14.A7 represents the proven petroleum reserves, which of course, depends on detection and exploration technology and related matters. In the row in brackets, we can see the share of each country's or region's petroleum reserves as a percentage of total world reserves. The numbers in brackets denote the percentage of the world total. In the second row the ratio of annual production (obtained by multiplying the second column by the number of days per year) to reserve are shown and in brackets the time to exhaustion.

As mentioned, if we take the total petroleum reserves in the United States, proven and unproven, at the current rate of extraction of petroleum through annual production, they will not last for more some decades. Taking world petroleum reserves as a whole, at current rates of production there are more than 40 years left before exhaustion. Considering individual countries at their present rates of production, for Iran there are 81 years, for Iraq 140, for Kuwait 150, for Saudi Arabia 96, and for the United Arab Emirates 86. Venezuela has 53 years to run.

The important point is that the Middle East petroleum resources— in particular those in Iraq—are very much underexploited. That is why the time to exhaustion in Iraq is very long. Table 14.A7 in fact underscores a major issue. On one hand, the reserves of the advanced economies are rapidly shrinking and thus overexploited, while on the other hand, those of the Middle East are both much larger and significantly underexploited.

As mentioned, the reserves consists of proved reserves and other reserves (inferred, measured, and indicated), which will be added to the proved reserves in the future. Of course, the exploration and production technology for petroleum is continuously changing, more rapidly than ever. This is the result of the so called DOFF—the "digital oil field of the future." The DOFF is based on information, exploration, and production far more exact and targeted. Recently, estimates of the recoverable reserves in Alberta, Canada, and Venezuela have been greatly increased by using new methods. The resources are now deemed recoverable because there have been great advances in the technology of processing tar and petroleum sands. Furthermore, new oil explorations, for example, in the Gulf of Mexico and Colombia, have come up with significant new findings.[16]

To sum up: in the United States, petroleum, natural gas, and coal together provide about 82 percent of all energy, and oil and gas roughly two thirds. Two thirds of the world's proved petroleum reserves are in the Middle East. In the United States, Mexico, Venezuela (less so in Canada), the remaining years to exhaustion of the known oil resources are only one fifth of the years to exhaustion in Middle East.

Substitutes for coal, natural gas, and crude petroleum are nuclear and renewable power. Because the share of these alternative energy production sources has increased in the past four decades, it seems likely that further research and development will enhance this trend. Also, an increased use of plastics particularly as a renewable substitute may relax the constraints that a sustainable development faces.

The foregoing assessments for nonrenewable resources are made under the assumption that only currently available technology is applied. But it seems to be likely that technological progress improves, for example, mining and refining methods or makes discoveries of new resources fields possible and therefore enhances present estimations of available reserves. Yet to ensure sustainable growth in the sense of minimal degradation of natural capital stocks and intergenerational equity, it appears to be important to develop further renewable substitutes. Given our rough assessment of the finiteness of the nonrenewable

---

[16] However, the costs of recovering petroleum from oil sands may prove daunting; almost as much energy must be used in processing as the final product will contain. Moreover, the processes generate severe pollution. The new strikes in the Gulf are in exceptionally deep waters and have already run into problems. Finally, all new unconventional petroleum will be expensive, and it will be some time before any comes on line. On the other hand, rapid advances in petroleum exploration and recovery are possible in all oil rich regions.

resources, we want to make an attempt to estimate a model with resource constraints.

14.6 ESTIMATION OF A BASIC MODEL

Here we introduce a methodological device of how to estimate a model with nonrenewable resources. To estimate the model as described in section 14.2, first consumer preferences and the technology of producing goods have to be specified, and second, a data set has to be constructed. We use data of a time period that is of particular interest, because the time series data cover the period of the large oil shocks in the 1970s. In our estimation, we focus on the standard model as presented in section 14.2.

We use a very simple version of a growth model with resources. It is assumed that the natural resource contributes to economic activity only in one way: it is used as an input factor for the production of a commodity that is either consumed or added to the capital stock. The present value of utility received from consumption is given by

$$\int_0^\infty \frac{C_t^{1-\eta} - 1}{1 - \eta} e^{-\delta t} \, dt,$$

where η is the elasticity of marginal utility. The technology of goods is described by a Cobb-Douglas-production function that depends on reproducible capital K_t and the exhaustible resource flow R_t. The elasticity of substitution between reproducible capital and the exhaustible resource σ equals 1. The evolution of capital is determined by

$$\dot{K}_t = K_t^\beta R_t^{1-\beta} - C_t.$$

β denotes the share of reproducible capital in production. Maximizing present value of utility and setting $y_t = C_t/R_t$ yields the following estimable system:[17]

$$\frac{\dot{y}_t}{y_t} = \frac{\beta x_t^{\beta-1} - \delta}{\eta} - \phi, \tag{14.16}$$

$$\frac{\dot{x}_t}{x_t} = x_t^{\beta-1}, \tag{14.17}$$

$$\frac{\dot{R}_t}{R_t} = \phi, \tag{14.18}$$

with ϕ as the growth rate of the exhaustible resource flow.

---

[17] For a detailed study of the solution, see appendix A.5.

Table 14.1 Estimation Results.

Parameter	Value	Standard Error
β	0.32	3.4668
$\bar{\delta}$	0.03	0.0369
$\bar{\eta}$	0.5	0.2488
ϕ	0.002	0.1856

For time series data we need consumption, reproducible capital stock,[18] and the exhaustible resource flow.[19]

The reproducible capital stock, K_t, is gross real private fixed capital stock, and C is private consumption. The time series for the resource, R, is based on our own computation. Because the total mineral production value is the amount of extracted exhaustible resources times average prices, it is used to measure the exhaustible resource flow. All time series are deflated by the GDP price index 1990 = 100.

Equations (14.16), (14.17), and (14.18) are estimated directly by using nonlinear least squares techniques (NLLS).[20] In the estimation we have prefixed the discount rate, $\bar{\delta}$, and the parameter of relative risk aversion, $\bar{\eta}$. The reason for this procedure is that the model we are considering—leaving aside substitution, technological change, and the role of other inputs—in its current form is rather incomplete, and reliable estimates for the discount rate as well as for relative risk aversion cannot be expected. Therefore, we prefix them at levels that have been obtained by other recent studies. The estimation results are summarized in table 14.1.

The estimated capital share in income, β, and the estimated growth rate of the resource flow, ϕ, are reasonable.

Although there is a very irregular behavior of the total mineral production value over the observed time period, the data show an enormous increase in the years 1972 to 1981 caused by the price effects of the oil crisis. Our estimation still gives reasonable parameter values.

Figures 14.1 and 14.2 show the estimated (fitted) and actual time series for the ratio of capital stock to resources and the ratio of consumption to resources.

As figures 14.1 and 14.2 show, the model with our estimated parameters matches the data well. As already noted, the very simple structure of the model may explain the observable slight correlation of the error terms. We also have supposed that population remains constant, which

[18] Data on consumption and capital stock are obtained from Citibase (1998).
[19] Source: U. S. Department of Commerce, Economics and Statistics (1965–1997).
[20] Computed with GAUSS OPTMUM version 3.00.

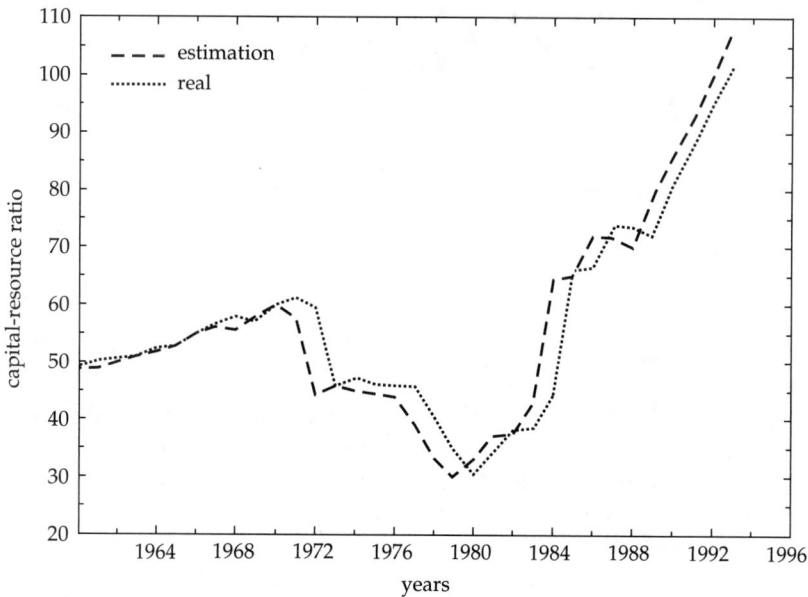

Figure 14.1 Capital Stock to Resource Ratio.

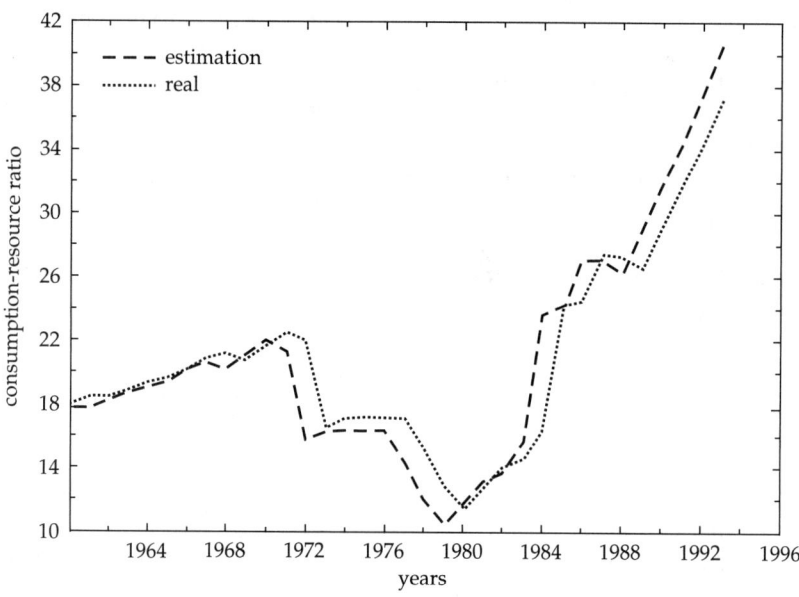

Figure 14.2 Consumption to Resource Ratio.

implies that labor as an input factor in production is constant, too. This assumption seems justified when analyzing the very short run, but not appropriate when examining a time period of 35 years. It seems to be likely that incorporating the factor labor into the model would improve the estimation results. Especially the exhaustible resource share in production, $1 - \beta$, is likely to be smaller. Furthermore, as the stylized facts support, it would be reasonable to allow technological change and substitution. Finally, the quality of nonlinear estimations depends strongly on the number of observations. Because the analyzed time period consist of only 35 data points, it is difficult to achieve sufficient robustness in the estimations. This problem was clearly observable when we attempted to estimate the prefixed parameters $\bar{\eta}$ and $\bar{\delta}$. Their estimation, in fact, turned out to be nonrobust with respect to the algorithm used. We therefore kept them prefixed.

In summary, future research should take into account the factor labor and technological progress and substitution and estimate such a model over a longer time period or with data with a higher frequency. Yet our methodology and estimation results for the period of the large oil crises in the 1970s and their effect on the 1980s and 1990s may already be of interest.

Appendix: Sketch of Solutions and Data Sources

A.1 THE BASIC MODEL

The following maximization problem must be solved:

$$Max \int_0^\infty U(C_t)e^{-\delta t}\, dt \tag{14.19}$$

subject to

$$\dot{K}_t = F(K_t, R_t) - C_t$$

$$\S_t = S_0 - \int_0^\infty R_t\, dt$$

$$\dot{S}_t = -R_t.$$

The current-value Hamiltonian is defined as

$$H_c = U(C_t) + p_t(-R_t) + q_t(F(K_t, R_t) - C_t), \tag{14.20}$$

where p_t denotes the shadow price of the resource constraint and q_t denotes the shadow price of capital accumulation. Computing the first order conditions yields the following equations, which must be fulfilled along an optimal path:

$$\dot{p}_t = \delta p_t \tag{14.21}$$

$$U'(C_t) = q_t \tag{14.22}$$

$$p_t = q_t F_R \tag{14.23}$$

$$\dot{q}_t - \delta q_t = -q_t F_K, \tag{14.24}$$

with $F_R = \partial F(K_t, R_t)/\partial R_t$ and $F_K = \partial F(K_t, R_t)/\partial K_t$. Differentiating (14.22) with respect to time and substituting (14.24) yields the consumption rate along an optimal path:

$$\frac{\dot{C}_t}{C_t} = \frac{F_K - \delta}{\eta}. \tag{14.25}$$

Differentiating (14.23) with respect to time and using (14.21) and (14.24) results in

$$F_K = \frac{\partial F_R}{\partial t} \frac{1}{F_R}. \tag{14.26}$$

Substituting $F_R = f(x_t) - x_t f'(x_t)$ and $F_K = f'(x_t)$ with $f(x_t) = F(K_t/R_t, 1)$ and $x_t = K_t/R_t$ in (14.26) yields the capital-resource ratio along an optimal path

$$\frac{\dot{x}_t}{x_t} = \sigma \frac{f(x_t)}{x_t}. \tag{14.27}$$

A.2 TECHNOLOGICAL CHANGE

The following maximization problem must be solved:

$$Max \int_0^\infty [U(C_t)\Omega_t + \omega_t W(K_t, S_t)]e^{-\delta t} \, dt \tag{14.28}$$

subject to

$$\dot{K}_t = F(K_t, R_t) - C_t$$

$$S_t = S_0 - \int_0^\infty R_t \, dt$$

$$\dot{S}_t = -R_t.$$

The current-value Hamiltonian is

$$H_c = U(C_t)\Omega_t + \omega_t W(K_t, S_t) + p_t(-R_t) + q_t(F(K_t, R_t) - C_t). \tag{14.29}$$

Computing the first order conditions yields

$$\dot{p}_t = \delta p_t \tag{14.30}$$

$$\Omega_t U'(C_t) = q_t \tag{14.31}$$

$$p_t = \omega_t W_S + q_t F_R \tag{14.32}$$

$$\dot{q}_t - \delta q_t = -\omega_t W_K - q_t F_K. \tag{14.33}$$

Differentiating (14.31) with respect to time and substituting (14.33) results in

$$\frac{\dot{C}_t}{C_t} = \frac{F_K - \delta + \psi_t(W_K - U'(C_t))/U'(C_t)}{\eta}, \tag{14.34}$$

with $\psi_t = \omega_t/\Omega_t$. If $W_k = W_S = 0$, one gets the following consumption rate along an optimal path

$$\frac{\dot{C}_t}{C_t} = \frac{F_K - (\delta + \psi_t)}{\eta}. \tag{14.35}$$

A.3 BACKSTOP TECHNOLOGY

The following maximization problem must be solved

$$Max \int_0^\infty U(C_t)e^{-\delta t}\, dt \tag{14.36}$$

subject to

$$\dot{K}_t = F(K_t, R_t) - C_t - g(z_t)R_t$$

$$S_t = S_0 - \int_0^\infty R_t\, dt$$

$$\dot{S}_t = -R_t,$$

where $g(z_t)R_t$ represents the total extraction costs. The current-value Hamiltonian is

$$H_c = U(C_t) + p_t(-R_t) + q_t(F(K_t, R_t) - C_t - g(z_t)R_t). \tag{14.37}$$

Computing the first order conditions yields

$$\dot{p}_t = \delta p_t + q_t g'(z_t)R_t \tag{14.38}$$

$$= \delta p_t q_t g'(z_t)\dot{z}_t$$

$$= \delta p_t q_t \frac{\partial g(z_t)}{\partial t}$$

$$U'(C_t) = q_t \tag{14.39}$$

$$-p_t = q_t(F_R - g(z_t)) \tag{14.40}$$

$$\dot{q}_t - \delta q_t = -q_t F_K. \tag{14.41}$$

Differentiating (14.40) with respect to time and using (14.38) and (14.41) results in

$$F_K = \frac{\partial F_R}{\partial t}\frac{1}{F_R} + \frac{F_K g(z_t)}{F_R}. \tag{14.42}$$

Substituting $F_R = f(x_t) - x_t f'(x_t)$ and $F_K = f'(x_t)$ in (14.42) results in the following capital-resource ratio along an optimal path

$$\frac{\dot{x}_t}{x_t} = \frac{\sigma f(x_t)}{x_t} + \frac{f'(x_t)}{x_t f''(x_t)}\frac{g(z_t)}{x_t}. \tag{14.43}$$

A.4 THE AMENITY VALUE OF A NATURAL RESOURCE

The following maximization problem must be solved

$$Max \int_0^\infty U(C_t, S_t)e^{-\delta t}\, dt \tag{14.44}$$

subject to

$$\dot{K}_t = F(K_t, R_t) - C_t$$

$$S_t = S_0 - \int_0^\infty R_t\, dt$$

$$\dot{S}_t = -R_t.$$

The current-value Hamiltonian is defined as

$$H_c = U(C_t, S_t) + p_t(-C_t) + q_t(F(K_t, R_t) - C_t). \tag{14.45}$$

Computing the first order conditions yields

$$\dot{p}_t - \delta p_t = -U_S \tag{14.46}$$

$$U_C = q_t \tag{14.47}$$

$$p_t = q_t F_R \tag{14.48}$$

$$\dot{q}_t - \delta q_t = -q_t F_K. \tag{14.49}$$

Differentiating (14.47) with respect to time and substituting (14.49) results in the following consumption rate along an optimal path:

$$\frac{\dot{C}_t}{C_t} = \frac{F_K - \delta - (U_{CS}/U_C)R_t}{\eta}. \tag{14.50}$$

Differentiating (48) with respect to time and using (14.46) and (14.49) yields

$$F_K = \frac{\partial F_R}{\partial t}\frac{1}{F_R} + \frac{U_S/U_C}{F_R}. \tag{14.51}$$

Substituting $F_R = f(x_t) - x_t f'(x_t)$ and $F_K = f'(x_t)$ in (14.51) yields the following capital-resource ratio

$$\frac{\dot{x}_t}{x_t} = \sigma\frac{f(x_t)}{x_t} + \frac{U_S/U_C}{x_t^2 f''(x_t)}. \tag{14.52}$$

A.5 ESTIMATION

The following maximization problem has to be solved:

$$Max \int_0^\infty \frac{C_t^{1-\eta}}{1-\eta}e^{-\delta t}\, dt \tag{14.53}$$

$$\text{subject to} \tag{14.54}$$

$$\dot{K}_t = K_t^\beta R_t^{1-\beta} - C_t \tag{14.55}$$

$$\dot{S}_t = -R_t. \tag{14.56}$$

The optimality conditions are

$$\frac{\dot{C}_t}{C_t} = \frac{F_K - \delta}{\eta}, \tag{14.57}$$

$$\frac{\dot{x}_t}{x_t} = \frac{f(x_t)}{x_t}. \tag{14.58}$$

With $F_K = \beta K_t^{\beta-1}R_t^{1-\beta} = \beta(K_t/R_T)^{\beta-1} = \beta x_t^{\beta-1}$ and $F(K_t/R_t, 1) = f(x_t) = x^\beta$, one gets

$$\frac{\dot{C}_t}{C_t} = \frac{\beta x_t^{\beta-1} - \delta}{\eta}, \tag{14.59}$$

$$\frac{\dot{x}_t}{x_t} = x_t^{\beta-1}. \tag{14.60}$$

Setting $y_t = C_t/R_t$ yields $\dot{y}_t/y_t = (\dot{C}_t/C_t) - (\dot{R}_t/R_t)$, and the following estimable system is obtained:

$$\frac{\dot{y}_t}{y_t} = \frac{\beta x_t^{\beta-1} - \delta}{\eta} - \phi \tag{14.61}$$

$$\frac{\dot{x}_t}{x_t} = x_t^{\beta-1} \tag{14.62}$$

$$\frac{\dot{R}_t}{R_t} = \phi, \tag{14.63}$$

where ϕ denotes the growth rate of the resource flow.

A.6 DATA SOURCES, FIGURES, AND TABLES

- Annual data from 1960 to 1995, taken from the U.S. Department of Commerce, Economics and Statistics, Bureau of the Census, *Statistical Abstract of the United States, 1965–1996*: crude petroleum production, coal production, natural gas production, copper production, iron ore production, zinc production, lead production, share of nuclear power, renewable energy and fuel minerals in total energy production, total mineral production value, proved reserves of crude petroleum and natural gas, gross private consumption, accumulated gross fixed capital formation.
- Estimated reserves of crude petroleum and natural gas, U.S. Geological Survey, *National Assessment of Oil and Gas Resources, 1995*: proved reserves, field growth, undiscovered resources, total reserves.
- Estimated reserves of coal, Energy Information Administration, *Coal Industry Annual 1996, Executive Summary 1996*: proved reserves, other reserves, total reserves.
- Estimated reserves of copper, iron ore, lead, and zinc, U.S. Geological Survey, *Mineral Commodity Summary, 1996*: proved reserves, other reserves, total reserves.

FIGURES

Fuel mineral production per 1990 $ GDP (coal production measured short tons, crude petroleum production measured in barrels, natural gas production measured in millions of cubic feet).

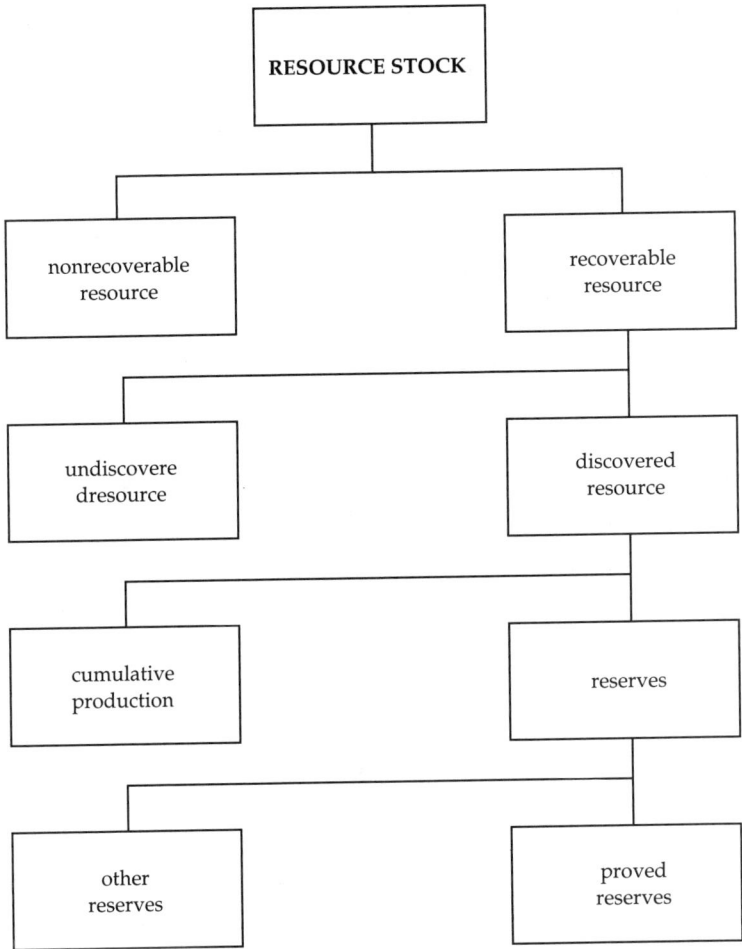

Figure 14.A1. The Components of a Resource Stock.

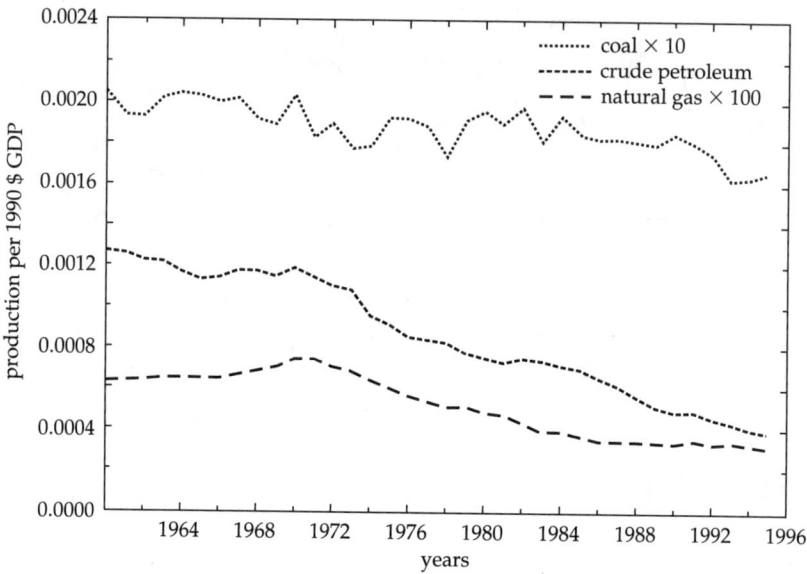

Figure 14.A2. Fuel.

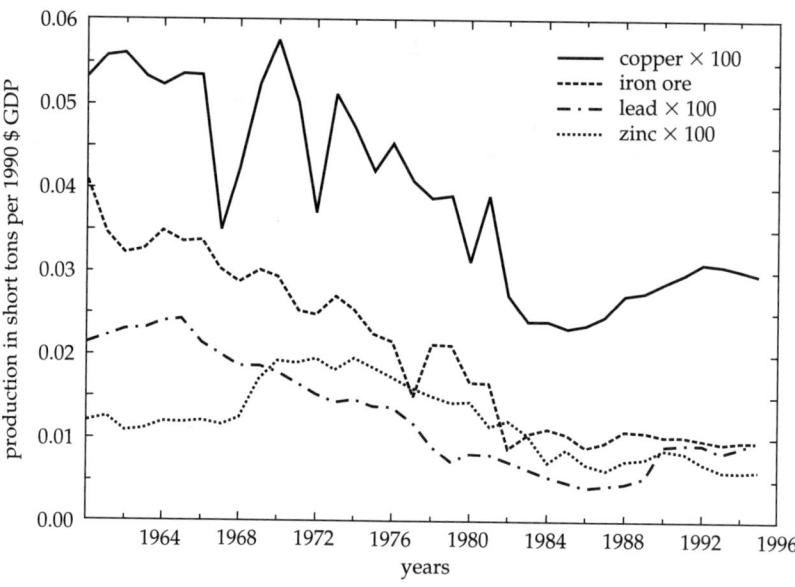

Figure 14.A3. Metal Production per GDP (metal production in short tons per 1990 $ GDP).

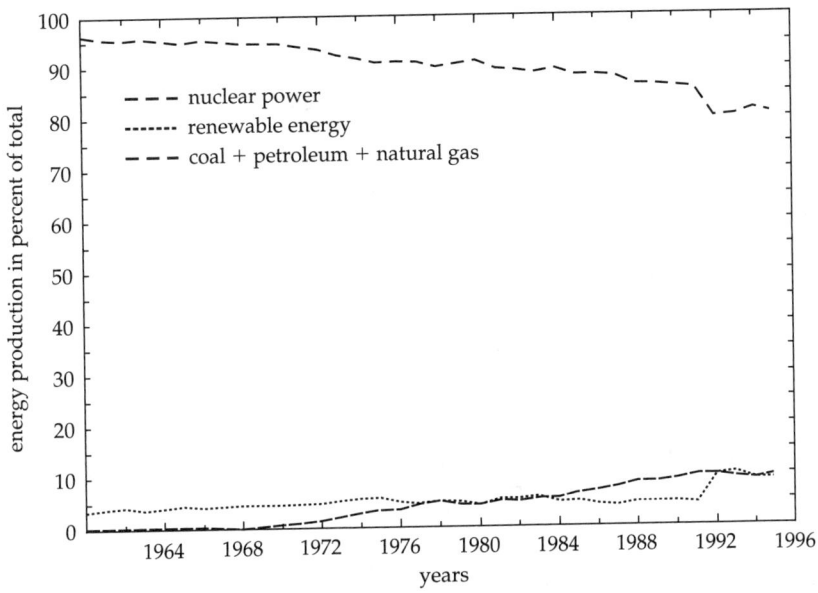

Figure 14.A4. Energy Production Sources in Percent of Total.

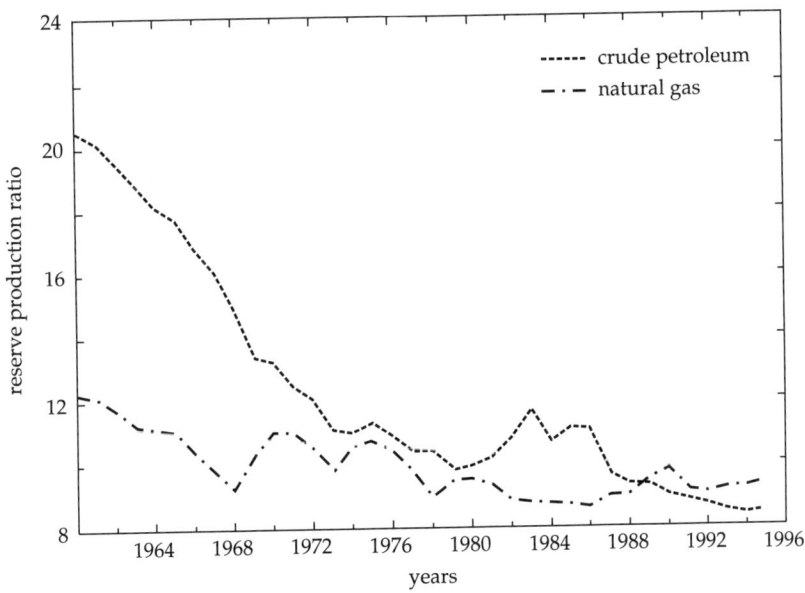

Figure 14.A5. Reserve-Production Ratio.

TABLES

Table 14.A1. Fuel Mineral Production per Dollar Real GDP.

	Crude Petroleum Barrels	Coal in Short Tons	Natural Gas in Millions of Cubic Feet
1960	$1.2680*10^{-3}$	$2.0526*10^{-4}$	$6.3155*10^{-6}$
1995	$3.8447*10^{-4}$	$1.6589*10^{-4}$	$3.1316*10^{-6}$
change in %	-70	-19	-48

Table 14.A2. Nonfuel Mineral Production per Dollar Real GDP.

	Copper in Short Tons	Iron Ore in Short Tons	Lead in Short Tons	Zinc in Short Tons
1960	$5.3287*10^{-4}$	$4.0952*10^{-2}$	$1.2187*10^{-4}$	$2.1463*10^{-4}$
1995	$2.9700*10^{-3}$	$9.8124*10^{-3}$	$6.1990*10^{-5}$	$9.8605*10^{-5}$
change in %	-44	-76	-49	-54

Table 14.A3. Estimated Reserves of Petroleum and Natural Gas, 1995.

	Crude Petroleum in Billion Bbl.	Natural Gas in Trillion. Cu. Ft.
Proved reserves	20.2	135
Field growth	60.0	322
Undiscovered resources	30.3	259
Total reserves	110.5	716
Average annual production/ average production growth rate	3.0076	0.012
Years left until exhaustion	40	300

Table 14.A4. Estimated Reserves of Coal 1996.

Proved Reserves	Other Reserves	Total Reserves	Average Production Growth Rate	Years Left Until Exhaustion
19,428	507,740	527,168	0.026	250

Table 14.A5. Estimated Reserves of Some Selected Metals, 1997.

	Copper in 1000 Metric Tons	Iron Ore in 1000 Metric Tons	Lead in 1000 Metric Tons	Zinc in 1000 Metric Tons
Proved reserves	45,000	10,000,000	6,500	25,000
Other reserves	90,000	23,000,000	20,000	80,000
Total reserves	135,000	33,000,000	26,500	105,000
Average annual production	1414.25	68844.44	439.72	437.14
Years left until exhaustion	95	480	60	240

Table 14.A6. Petroleum Import Shares.

	Persian Gulf/Middle East	Iran	Iraq	Kuwait	Saudi Arabia	United Arab Emirates	Venezuela
U.S. (% of imports from)	2.25 [20.20]	0 [0]	0.32 [2.88]	0.22 [1.98]	1.68 [15.11]	0.012 [0.11]	1.32 [11.87]
OECD Europe (% of imports from)	2.56 [35.96]	0.74 [10.39]	0.58 [8.15]	0.09 [1.26]	1.13 [15.87]	0.04 [0.56]	0 [0]

Table 14.A7. Petroleum Production and Petroleum Reserves.

	World Total	OPEC	Persian Gulf/ Middle East	Iran	Iraq	Kuwait	Saudia Arabia	Unit. Arab Emirates	Venezuela	United States	Russia
Reserves (2001) (billion barrels)[% of world total]	1.017,73		662,48 [65.09]	99.08 [9.74]	115,00 [11.3]	98,85 [9.71]	261,65 [25.71]	62,82 [6.17]	50,22 [4.93]	21,50 [2.11]	53,86 [5.29]
Ratio of production to reserves (2001) (annual production/ reserves)	0.024 [41.6]			0.0123 [81.3]	0.00073 [140.1]	0.00067 [149.1]	0.0104 [96.1]	0.0117 [85.5]	0.0189 [52.9]	0.057 [17.5]	0.047 [21.0]

15

Renewable Resources

15.1 INTRODUCTION

Renewable resources are empirically more difficult to treat.[1] Whereas stocks of nonrenewable resources can only be estimated with great uncertainty, the estimation problem is compounded for renewable resources. Here we only present some theoretical models. Yet what we take into account more explicitly is the interaction of resources.

For a long time resource economists, applying dynamic optimization, have confined their studies on the depletion of renewable resources to the analysis of a single isolated resource.[2] Recent work in resource economics has embarked on studying the exploitation of ecologically interrelated resources. The optimal exploitation of interrelated resources is frequently modeled as an optimally controlled Lotka-Volterra system.[3] It might, however, be fair to state the basic infinite horizon discounted optimal control problem has not completely been resolved since it was first introduced in 1976 by Clark (1990). Yet there are numerous partial models and also precise conjectures—due to Clark (1990) and Ragozin and Brown (1985)—about the type of solutions that one would expect; see section 15.2.

In this chapter, we make use of some advancement in analytical methods and employ computer studies to discuss a number of issues and eventually disprove—by means of counterexamples—some statements made in the literature on optimal exploitation of interacting resources. In the study of the fate of optimally exploited interacting resources, there are essential modeling issues involved. The following issues have become important and are considered in our various models: (1) zero time horizon optimization and infinite time horizon optimization, (2) sole owner control over harvesting and free access to harvesting, (3) selective (independent harvesting) and nonselective harvesting (joint harvesting) of different resources, (4) elastic and inelastic

---

[1] This chapter relies heavily on the joint work of Willi Semmler with Malte Sieveking.
[2] This tradition started with the pathbreaking work by Hotelling (1931). For a recent survey, see Heal (1999).
[3] See, for example, Clark (1990, 1985), Hannesson (1983), Ragozin and Brown (1985), Conrad and Adu-Asamoah (1986), and Falk (1988). A detailed discussion of this literature is undertaken in section 15.2.

demand curves, and (5) predator-prey and competitive interaction of resources.

It can be shown that zero time horizon optimization is related to the infinite time horizon optimization by a limit process in which the discount rate tends to infinity and reduces that latter to a simple differential equation. We provide a counterexample to Clark's (1985) fundamental principle of renewable resources, which says that higher discount rates normally imply a lower level of resource conservation by private owners, other things being equal. As also shown, the sole owner, who has been said to be more conservative with respect to the exploitation of the resource than the free access regime because of his capability to plan, may turn out to be of more serious threat to extinction than free access in some cases. The same appears to hold true for selective and nonselective harvesting. The former is often considered a sharper strategy, yet as will be seen, it may be more dangerous in certain situations. We also show that inelastic demand may cause the resources to oscillate and produce limit cycles when optimally exploited, whereas the more elastic demand may lead to stable equilibria. Oscillations—limit cycles—will also be shown to occur in a system of competing species that without being harvested, never oscillates.

The foregoing results are generated partly analytically, partly by means of numerical studies. Because of the possible nonconcavity of the Hamiltonian involved in our optimization problems, we do not apply Pontryagin's maximum principle[4] but rather study the dynamics of the optimally controlled system by means of a dynamic programming algorithm. The numerical studies are undertaken for specific examples, specific functional forms,[5] and parameters. Fully worked out examples are given for two resources that either exhibit predator-prey or competitive interrelations. Although our specifications and simulations may not have the same generic status as analytical statements, they are, however, indispensable in tracking the exact long-term behavior of the trajectories of the optimally controlled systems.

The remainder of this chapter is organized as follows. Section 15.2 introduces the model and various specifications of it as well as two theorems. The theorems reveal the relation between zero horizon

---

[4] A dynamic programming algorithm is preferred in our context because it approximates truly optimal trajectories as compared to procedures working on the basis of necessary conditions like Pontryagin's maximum principle. The latter might only yield "suboptimal" trajectories. This is the reason the Hamiltonian method employed, for example, by Dockner and Feichtinger (1991) who show the existence of optimal periodic solutions for certain variants of their model (under the assumption of a concave Hamiltonian) is difficult to apply in our context.

[5] A variety of different functional forms, for example, concerning the inverse demand function, are examined in Sieveking and Semmler (1990) with similar outcomes for the resource dynamics as reported here in.

optimization where the discount rate tends to infinity, and infinite time horizon optimization, where the discount rate is finite. A brief discussion on the related literature is added. In section 15.3, analytical results for the open access regime—in the literature typically viewed as a zero horizon optimization problem—are gathered concerning the systems of predator-prey and competitive interactions. Both interactions allow for limit cycles in the optimal trajectories. Numerical studies are added to support and extend the results. Section 15.4 presents numerical results for the monopoly problem, by nature an infinite horizon optimization problem. To track the exact solution paths of both systems for varying discount rates—possibly exhibiting limit cycles, instability, or extinction—a dynamic programming algorithm is employed.[6]

Finally, we note that renewable resources can be considered (at least to a great extent) as environmental assets. We have set up a typical dynamic resource exploitation model, with a discount payoff as objective function, so that the objective function gives us the present value of the resource (or resources). This way we can also study the issue of pricing of environmental assets. We study this issue in a deterministic framework and do not consider a stochastic modeling of the pricing of environmental assets, though the algorithm applied here may help do this.[7]

15.2 RENEWABLE RESOURCES AND OPTIMAL GROWTH

Following the aforementioned bioeconomic literature, we commence with simple Lotka-Volterra equations for two resources, for example, two species x_1, x_2 that exhibit intraspecific competition:

$$\dot{x}_1 = x_1(a_0 - a_1 x_1 - a_2 x_2) = f(x_1.x_2),$$
$$\dot{x}_2 = x_2(b_0 + (-)b_1 x_1 - b_2 x_2) = f(x_1, x_2), \tag{15.1}$$

where a_i, b_i are positive constants, except b_0, which may also be a negative constant. The terms $-a_1 x_1$ and $-b_2 x_2$, respectively, denote intraspecific competition, for example, population pressure. In taking $-a_1 x_1, -b_1 x_1$ we wish to represent two species that compete for the same resources, while $-a_2 x_2, +b_1 x_1$ stands for x_1 being the prey for the predator x_2.

If we take harvesting into account, we do this by writing

$$\dot{x}_1 = f(x_1, x_2) - x_1 u_1,$$
$$\dot{x}_2 = f(x_1.x_2) - x_2 u_2, \tag{15.2}$$

---

[6] The dynamic programming algorithm is described in Semmler and Sieveking (2000); see also Falcone (1987). A further development and refinement of this algorithm has been undertaken in Grüne and Semmler (2004).

[7] See Grüne and Semmler (2004).

where $u_1 \geq 0, u_2 \geq 0$ denote harvesting effort to be chosen by the economic agents. We assume the yield y_i of the effort u_i at the stock level x_i to be given by

$$y_i = (x, u) = q_i x_i u_i, \tag{15.3}$$

where q_i is a nonnegative constant.[8] Other functional relationships than (15.2) and (15.3) between effort rate, growth rate, and yield have been discussed (see Clark 1985). We choose the simplest version.

The incentives for the agent to choose a specific u_i is to increase the total net income flow

$$\sum_1^2 p_i(y_i) y_i - c_i(u_i) = G(x, u), \tag{15.4}$$

where $p_i(y_i)$ is the price per unit of biomass of x_i depending on the rate of supply y_i, and $c_i(u_i)$ is the cost per unit of effort of type i. The subsequent formula expresses the fact that our agent has complete control over the harvesting process. We therefore call him or her sole owner. The price of the harvested resource is determined by an inverse demand function. We assume that $p_i(y)$ has the shape

$$p_i(y_i) = (\gamma_0 + y_i)^{-\alpha} \tag{15.5}$$

for some $\alpha > 0$, $\gamma_0 > 0$. The reason for this choice is that we want to posit that

$$p_i > 0, \; p_i' > 0, \; p_i'' = 0; \qquad \lim_{y_i \to \infty} p_i(y_i) < +\infty.$$

These properties permit a simple analytical study of the dynamics.[9]

Note that we may equivalently view the sole owner as an agent who fixes a price and then satisfies the resulting demand. Moreover, for simplicity we take cost

$$c_i(u_i) = c_i u_i \tag{15.6}$$

for some constant $c_i \geq 0$.

The set U of admissible control vectors (u_1, u_2) is capable of representing characteristics of the harvesting technology as follows. We define *selective harvesting* by

$$U = \{(u_1, u_2) \mid 0 \leq u_1 \leq \bar{u}_1, \; 0 \leq u_2 \leq \bar{u}_2\}, \tag{15.7}$$

---

[8] In later computations and computer simulations we will set $q = 1$.
[9] Other forms of inverse demand functions are explored in Sieveking and Semmler (1990).

where $\bar{u}_1, \bar{u}_2$ are positive constants or $+\infty$. Subsequently, in the analysis and numerical simulations of the model, we confine the study to the case where only one resource is permitted to be caught (or salable), but the other resource is caught as well.[10] This is called *nonselective harvesting* and is represented by

$$U = \{(u_1, u_2) \mid 0 \leq u_2 \leq \bar{u}_2, = k_2 u_2\}, \tag{15.8}$$

where k_2 is a non-negative constant.

There are two problems connected with the sole owner that we wish to study: the infinite horizon problem and the finite horizon problem. The infinite horizon problem can be described as

$$\max \int_0^\infty e^{-\delta t} G(x(t), u(t)) dt \tag{15.9}$$

s.t. (2) and

$$u(t) = (u_1(t), u_2(t)) \in U(t \geq 0); \quad x(0) = x_0.$$

Here again, δ is the discount rate, which is likely to play an important role in models of resource management. According to the following theorem there is a limit in (15.9) as δ tends to infinity.

Theorem 15.2.1 *Let (x_n, u_n) be a solution[11] to (9) for discount rates*

$$\delta_n (n \in \mathbb{N} \text{ such that } \lim_{n \to \infty} \delta_n = +\infty \quad \text{and} \quad \lim_{n \to \infty} x_n(0) = a.$$

Then a subsequence of $(x_n)_{n \in \mathbb{N}}$ converges uniformly on bounded intervals of $[0, +\infty)$ to a solution of the following differential inclusion

$$\dot{y}(t) \in F(y(t)); \quad y(0) = a, \tag{15.10}$$

where

$$y(t) = (y_1(t), y_2(t)),$$

and

$$f(y_1, y_2) = \{(f_1(y_1, y_2) - y_1 u_1, f_2(y_1, y_2) - y_2 u_2) \mid\mid u = (u_1, u_2) \in U$$

and

$$G(y, u) = \max G(y, U)\}.$$

---

[10] For details of such models, see Clark (1985).
[11] For details of the subsequent theorem, see Semmler and Sieveking (1994b).

Equation (15.10) has been called a free access problem in Falk (1988), but we prefer to call it differently. In a free access regime there is unrestricted entry, and the coordination of activities is not feasible so as to maximize $G(x, u)$. We therefore call (15.10) the *zero horizon optimization problem* for the sole owner (monopolist).[12] It will turn out that for our particular f_1, f_2, and G, (15.10) is actually a system of ordinary differential equation

$$\dot{x}_1 = f(x_1, x_2) - x_1 u_1(x),$$
$$\dot{x}_2 = f(x_1, x_2) - x_2 u_2(x),$$

(15.11)

with $u(x)$ uniquely determined by

$$G(x, u) = \max G(x, U).$$

In fact we will have to consider only

$$u_2(x) = u_2(x_2) \quad \text{and} \quad u_1(x) = k_2 u_2(x_2)$$

for some constant $k_2 \geq 0$.

Our concept of free access regime generalizes Gordon's (1954) idea of dissipation of economic rent from the stationary state (Gordon's case) to any state at any time. That is, we posit agents to enter the industry as long as there is $G(xu) > 0$, thereby increasing the total effort $U = (u_1, u_2)$ until eventually $G(x, u) = 0$. If the adjustment process that drives rent to zero is infinitely fast, we arrive at the following system, which we call *open access problem*:

$$\dot{x}_1 = f(x_1, x_2) - x_1 v_1(x),$$
$$\dot{x}_2 = f(x_1, x_2) - x_2 v_2(x),$$

(15.12)

where $v(x) = (v_1(x), v_2(x))$ is determined by

$$v(x) = \sup\{u \in U \mid G(x(x, u) \geq 0\},$$

and U is partially ordered by

$$(u_1, u_2) \leq (u_1^*, u_2^*) \Longleftrightarrow u_1 \leq u_1^* \quad \text{and} \quad u_2 \leq u_2^*.$$

In general $v(x)$ will be a set, and hence (15.12) is to be read as differential inclusion. In the following examples, however, v is single-valued:

$$v_2(x) = v_2(x_2) \quad \text{and} \quad v_1(x) = k_2 v_2(x_2),$$

---

[12] Although, as some authors have pointed out, the monopolist's problem is by nature an infinite horizon optimization problem, we employ the limit case as a modeling device to obtain conjectures for the infinite horizon case.

for some $k_2 \geq 0$ and hence (15.12) is a system of two ordinary differential equations. We also term (15.12) a zero time horizon problem. For a model where the adaptation of effort is modeled by a differential equation, see Clark (1990, p. 322).

The zero horizon optimization problem is a helpful modeling device for studying the infinite horizon case. In fact, because the former represents a system of ordinary differential equations in the plane, we are able to study the existence of limit cycles via Hopf bifurcation theory or the Poincaré-Bendixson theorem.[13] From the horizon optimization problem, one may conclude back to the infinite horizon problem according to theorem 15.2.2.

Theorem 15.2.2 *Suppose the sole owner zero horizon problem (15.11) admits a stable limit cycle; then for sufficiently large $\delta > 0$, problem (15.9) admits a limit cycle solution, too.*

Theorem 15.2.2 follows from a general theorem according to which limit cycles and asymptotically stable attractors are structurally stable. The theorem is, of course, also applicable to other parameters than δ (for details, see Sieveking 1990). Thus, the study of the zero horizon problem and theorem 15.2.2 guided us in the search for specific parameters for the simulations of the infinite horizon problem in section 15.4.

It is worth contrasting the properties of our system (15.9) with the literature in bioeconomics and the turnpike solution in optimal growth theory. In bioeconomic literature such as Clark (1985, 1990), there is open access or sole owner (who faces an inverse demand function) in the case of a single resource. In case of interacting resources the literature, to our knowledge, assumes constant sales prices. Yet no general solution has been given (see Clark 1985, p. 175) and there has not been significant progress since 1976 (Clark 1990).

To fix ideas, let us consider the following problem

$$\max \int_0^\infty e^{-\delta t}(px_2(t) - c)u_2(t)dt$$

$$\text{s.t. } \dot{x}_1 = x_1(a_0 - a_1x_1 - a_2x_2) = f_1(x_1, x_2),$$

$$\dot{x}_2 = x_2(b_0 - b_1x_1 - b_2x_2) = f_2(x_1, x_2) - u_2x_2,$$

$$0 \leq u_2 \leq \bar{u}_2,$$

$$x(0) = x_0. \tag{15.13}$$

---

[13] There is, however, also an economic reason for considering the zero horizon problem. Clark, for example, notes that "resource ownership forces the exploiter to adopt in essence an infinite discount rate" (Clark 1985, p. vi). Note, however, that we do not identify the case of an infinite discount rate with an open access regime. These appear to be two distinct cases.

Here $p, c, a_0, a_1, a_2, b_1, b_2$ are positive constants and $b_0 \in \mathbb{R}$, that is, we consider selective harvesting of a predator by a sole owner who faces a fixed price.

The current-value Hamiltonian of (15.13) is defined to be

$$H = \lambda_0 (px_2 - c)u_2 + \lambda_1 f_1(x_1, x_2) + \lambda_2 (f_2(x_1, x_2) - x_2 u_2).$$

According to Pontryagin's maximum principle (MP), if (x, u) is a solution to (15.13), there exists $\lambda(t) = (\lambda_0, \lambda_1(t), \lambda_2(t))$ such that

$$\lambda_0 \in \{0, 1\}, \quad \lambda(t) \neq 0 \quad \text{for all } t \geq 0,$$

$$H(x(t), \lambda(t), u(t)) = \max\{H(x(t), \lambda(t), v) \mid 0 \leq v \leq \bar{u}_2\}$$

and

$$\begin{aligned}
\dot{\lambda}_1 &= \delta\lambda_1(t) - \frac{\partial}{\partial x_1} H(x(t), \lambda(t), u(t)), \\
\dot{\lambda}_2 &= \delta\lambda_2(t) - \frac{\partial}{\partial x_2} H(x(t), \lambda(t), u(t)).
\end{aligned} \tag{15.14}$$

The latter are the adjoint equations. Let us call[14] a solution $(x(t), u(t))$ to the necessary conditions of the MP *bang-bang* on an interval I, if for all $t \in I$, $u(t) = \{0, \bar{u}_s\}$ and *singular* on I, if for all $t \in I$,

$$\lambda_0 (px_2 - c) - \lambda_2 x_2 = 0.$$

Clark (1990) and Ragozin and Brown (1985) conjecture the following. *Turnpike Property*: There exists a unique stationary solution (x_∞, u_∞) to (15.13) that is singular on $[0, \infty)$ and such that for every solution $(x(t), u(t))$ to (13),

$$\lim_{t \to \infty} x(t) = x_\infty.$$

The proviso has to be made that the stationary states with $u_2 = 0$ and $u_2 = \bar{u}_2$, respectively, are not optimal. In fact, these authors noted that there exists a unique stationary singular solution (x_∞, u_∞) to the MP conditions and a unique pair $(x^*, u^*), (x^{**}, u^{**})$ of singular solutions to the MP conditions such that

$$\lim_{t \to \infty} x^*(t) = x_\infty = \lim_{t \to \infty} x^{**}(t),$$

and $[0, +\infty)^2$ is divided into connected components by these solutions. They conjectured that any solution to (15.13) is steered in a bang-bang

---

[14] For the following statement, see also Ragozin and Brown (1985).

fashion (most rapidly) to the stable manifold, which is composed of the trajectories x^* and x^{**} and the equilibrium x_∞.

A proof of the turnpike property of predator-prey systems (15.13) and perhaps some systems of competitive interactions would probably make an important contribution to optimal control theory. However, a numerical test with several sets of randomly chosen parameters and a dynamic programming algorithm could already be fairly convincing.

There is an important literature where the turnpike property is proved, for example, Rockafellar (1976), Brock and Scheinkman (1976), Cass and Shell (1976), Carlson and Haurie (1987), and Sorger (1989). These authors, however, assume that the Hamiltonain H is concave jointly in u and x. In that case, with small discount rates an asymptotically stable optimal path is obtained; see Cass and Shell (1976), Brock and Scheinkman (1976), and Rockafellar (1976). For large discount rates, as demonstrated by Boldrin and Montrucchio (1986), irregular (chaotic) dynamic behaviors are admissible solutions in optimal growth theory. Under the hypotheses of a concave Hamiltonian H, other studies— such as Benhabib and Nishimura (1979) and more recently Dockner and Feichtinger (1991)—demonstrate that limit cycles exist. The demonstration is undertaken via Hopf bifurcation analysis for the system of state equations plus adjoint equations with control u eliminated via the MP.

In our case, the Hessian of H with respect to x and u is

$$\begin{bmatrix} \lambda_1 \dfrac{\partial^2}{\partial x_2^2} f_1 + \lambda_2 \dfrac{\partial^2}{\partial x_2^2} f_2 & \lambda_0 p - \lambda_2 \\ \lambda_0 p - \lambda_2 & 0 \end{bmatrix}.$$

The characteristic polynomial has a positive and negative root, which shows that H is not concave jointly in x and u. Therefore, because the Hamiltonian is not of concave type, the cited literature on optimal growth theory does not apply directly to the problems we consider here.[15]

We need to add, however, that in the present article our intention is not to solve the aforementioned turnpike problem; rather, we show by examples that the turnpike property may fail to hold if prices are allowed to be nonconstant.

---

[15] We want to note, however, that optimal growth theory has also departed from the assumption of a concave Hamiltonian. Based on original work of Clark (1971), a nonconvex production technology is posited; see Majumdar and Mitra (1982), and Dechert and Nishimura (1983). These publications show that for an economy with nonconvex technology (i.e., with first increasing then decreasing returns to scale) a sufficiently high δ will give rise to a depletion of capital stock, causing the economy's extinction.

15.3 OPEN ACCESS: ZERO HORIZON OPTIMIZATION

As mentioned, we characterize the open access exploitation as zero horizon optimization problem. In theorem 15.3.1 we obtain the shape of the optimal effort depending on the elasticity of demand. Limit cycles for the predator-prey system and the system of competing species are established in theorems 15.3.2 and 15.3.3. The detected limit cycles are then replicated in a numerical study. Let us first consider the impact of the elasticity of demand in the optimal effort.

15.3.1 Elasticity of Demand and Optimal Effort

We consider the effort $v_\alpha(x)$ in the open access case (15.12), which by definition is determined by

$$v_\alpha = \max\{v \mid p_\alpha(xv)xv - cv \geq 0; \, v \geq 0\},$$

$$p_\alpha(y) = (\gamma_0 + y)^{-\alpha}, \tag{15.15}$$

where (x, v) stands for (x_2, v_2). Thus, only one effort, v_2, is spent. Although the first resource is harvested as a byproduct, it is not salable. The second resource generates a yield $y = x_2 v_2$.

Theorem 15.3.1 (*open access optimal effort*):

a. *There is a unique solution $v_\alpha(x)$ to (15.15)*:

$$v_\alpha(x) = 0 \quad \text{for } 0 \leq x \leq \gamma_0^\alpha \text{ ß,}$$

and

$$v_\alpha(x) = c^{-1/\alpha} x^{1/\alpha - 1} - \gamma_0 x^{-1} \text{ for } \gamma_0^\alpha c < x.$$

b.
$$v_\alpha'(x) > 0 \Longleftrightarrow 0 < \alpha \leq 1 \quad \text{or} \quad x < \gamma_0^\alpha c \left(\frac{\alpha}{\alpha - 1}\right)^\alpha.$$

c. *The optimal supply $x v_\alpha(x)$ is increasing in x.*

For a derivation of these results, see appendix A.1.

15.3.2 Predator-Prey System

Consider a predator-prey system, such as

$$\dot{x} = x_1(a_0 - a_1 x_1 - a_2 x_2 - v_1(x_1)),$$
$$\dot{x} = x_2(-b_0 - b_1 x_1 - b_2 x_2 - v_2(x_2)), \tag{15.16}$$

where a_i, b_j are positive and v_1, v_2 non-negative but not necessarily optimal effort functions. We assume that (15.16) admits a unique stationary state $(\bar{x}_1, \bar{x}_2)$ in $(0, +\infty)^2$:

$$\bar{x}_1 > 0, \quad \bar{x}_2 > 0,$$
$$a_0 - a_1 \bar{x}_1 - a_2 \bar{x}_2 - v_1(\bar{x}_1) = 0,$$
$$-b_0 - b_1 \bar{x}_1 - b_2 \bar{x}_2 - v_2(\bar{x}_2) = 0.$$

We shall call the function $v : (0, +\infty) \to \mathbb{R}$ increasing at $\bar{x}_i (i = 1, 2)$ if $v(x') < v(\bar{x}_i) < v(x'')$ for $0 \leq x' \leq \bar{x}_i \leq x'' < +\infty$, and locally decreasing at $\bar{x}_i (i = 1, 2)$, if there exists $\epsilon > 0$ such that $v(x') > v(\bar{x}_i) > v(x'')$ for $\bar{x}_i - \epsilon < x' < \bar{x}_i < x'' < \bar{x}_i + \epsilon$.

Theorem 15.3.2 (*limit cycles for the predator-prey system*):

a. *Suppose $a_1 x_1 v_1(x_1)$ and $b_2 x_2 + v_2(x_2)$ are increasing at $\bar{x}_1$ and $\bar{x}_2$, respectively. Then $(\bar{x}_1, \bar{x}_2)$ is globally asymptotically stable for $(0, +\infty)^2$; thus every trajectory with positive coordinates spirals toward $(\bar{x}_1, 2)$.*

b. *Suppose $a_1 x_1 v_1(x_1)$ and $b_2 x_2 + v_2(x_2)$ is locally decreasing at $\bar{x}_1$ and $\bar{x}_2$, respectively. Then any trajectory that starts sufficiently close to $(\bar{x}_1, \bar{x}_2)$ but is nonstationary spirals toward a limit cycle.*

Note that the condition of a is satisfied if v_i both are strictly increasing and the condition of b is satisfied if $a_1 + v_1'(\bar{x}_1) < 0, b_2 + v_2'(\bar{x}_2) < 0$. Thus, different dynamics are generated via the variation of the elasticity α. For the proof of theorem 15.3.2, see appendix A.3.

15.3.3 System of Competing Species

Generically, every trajectory of a system of two species that are competing intraspecifically as well as interspecifically will converge to some equilibrium as time t tends to infinity; see Hirsch and Smale (1974). It is therefore interesting to note that such a system may begin to oscillate and even tend to a limit cycle when exploited. The system we study is

$$\dot{x}_1 = x_1(a_0 - a_1 x_1 - a_2 x_2 - v_1(x_1)),$$
$$\dot{x}_2 = x_2(b_0 - b_1 x_1 - b_2 x_2 - k_1 v_1(x_1)).$$

Here the first species is harvested, and the harvesting technology is unable to select between x_1 and x_2. The system exhibits multiple equilibria, which are studied in appendix A.4. A particular case, exhibiting three equilibria, is depicted in figure 15.1.

Let $\sigma^+(\sigma^-)$ be the trajectory that emanates from (tends to) x with increasing (decreasing) first coordinates. Similarly let $\tau^+(\tau^-)$ be the trajectory emanating from (tends to) y with first coordinates decreasing (increasing).

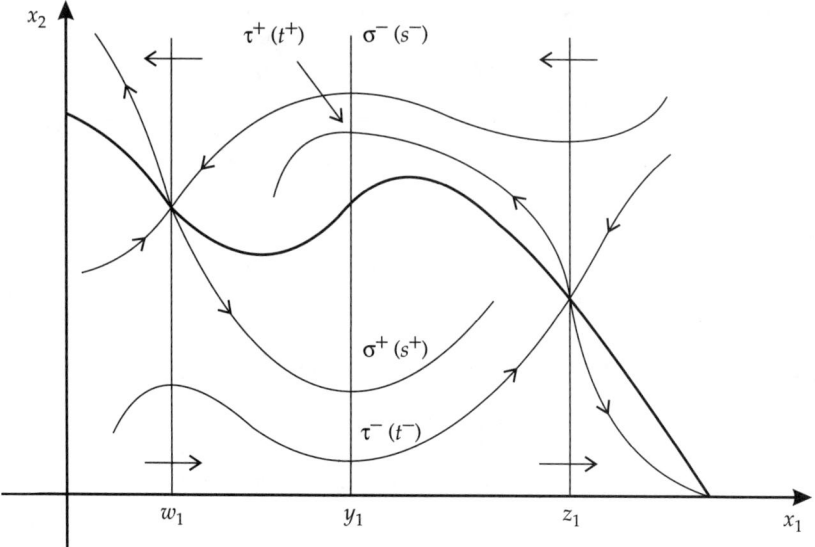

Figure 15.1 Phase Portrayal for the System of Competing Species.

Let $s^+(s^-)$ be the first instant t such that

$$\sigma^+(t)_1 = y_1 \quad \text{and} \quad \sigma^-(t)_1 = y_1, \text{ respectively.}$$

Similarly, $t^+(t^-)$ denotes the first instant t such that

$$\tau^+(t)_t = y_1 \quad \text{and} \quad \tau^-(t)_t = y_1, \text{ respectively.}$$

The reader may thus convince him or herself that figure 15.1 illustrates the following theorem.

Theorem 15.3.3 *(limit cycles for the system of competing species): If* $\sigma^+(s^+)_2 > \tau^-(t^-)_2$, *and* $\tau^+(t^+)_2 < \sigma^-(s^-)_2$ *then a limit cycle around* y *exists.*

In fact there are four isoclines and eight separatrices, two of which tend to a limit cycle. Trajectories below τ^- and to the left of x_1 will tend to some $(x^+, 0)$. Those above τ^+ and to the right of w_1 will tend to some $(0, y^*)$ where $y^* < +\infty$ if $b_2 > 0$.

Let us now turn to the numerical study.[16] For the predator-prey system the following parameters are employed:

$$a_0 = 1.04, \ a_1 = 0.001, \ a_2 = 0.07, \ b_0 = 1.01, \ b_1 = 0.2,$$
$$b_2 = 0.001, \ c = 0.25, \ k_1 = 1, \ y_0 = 1.$$

---

[16] The subsequent parameters possibly generating limit cycles were obtained through Hopf bifurcation analysis by Harald Diefenbach.

For the system of competing species the parameters employed are

$$a_0 = 3, \ a_1 = 0.122, \ a_2 = 0.75, \ b_0 = 2.251, \ b_1 = 0.0625,$$
$$b_2 = 0.001, \ c = 0.25, \ k_1 = 1, \ \gamma_0 = 1.$$

The predator-prey system, in the open access case, $\alpha = 2$, creates locally unstable trajectories. In fact, a limit cycle arises as predicted from theorem 15.3.2b; see figure 15.2.
Thus, the fate of resources depends on the elasticity of demand.[17]
The result for the system of competing species are as follows. First, the open access exploitation appears likely to drive a resource to extinction when α is small. Figure 15.3 illustrates this for $\alpha = 1$.

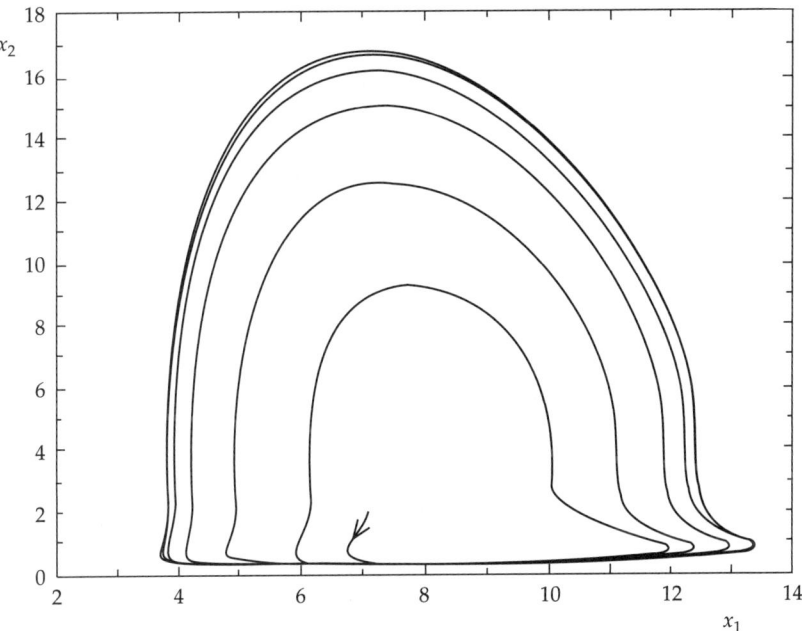

Figure 15.2 Predator-Prey, Open Access, $\alpha = 2$, Initial Values (7,2).

[17] Note that for the monopoly case with a discount rate tending to infinity, there are limit cycles predicted for $\alpha = 1$; see appendix A.2. This was replicated by simulations that also have shown that a steeper demand curve, that is, a greater α creates larger cycles and the resources tended to be extinct. The result that the monopoly case is less conservative than the open access case for the predator-prey system contradicts the usual statement for one-dimensional resource examples where the sole owner appears to be more conservative (see Clark 1990, ch. 2.5). Note that according to the theorem 15.2.2 our results must also hold in the monopoly case for finite but large discount rates.

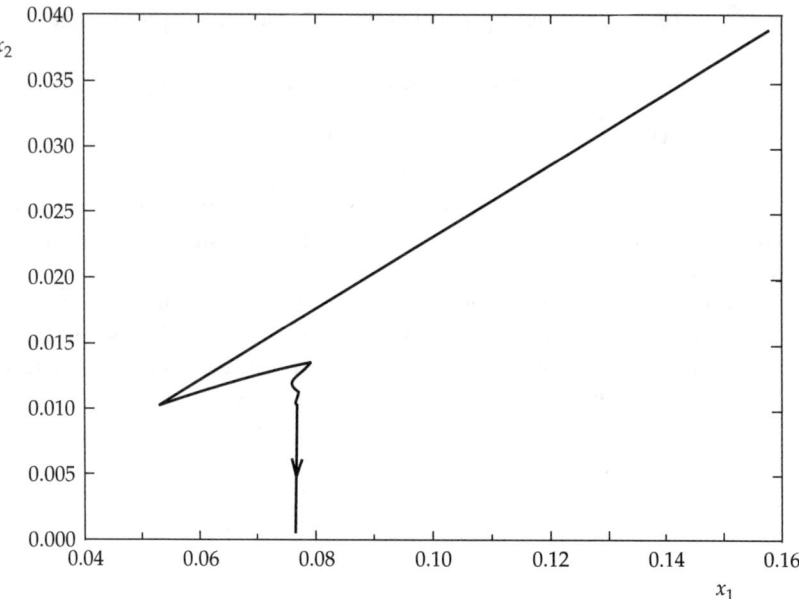

Figure 15.3 System of Competing Species, Open Access, $\alpha = 1$, Initial Values (0.016,0.004).

With a less elastic demand, $\alpha = 2$, a limit cycle[18] arises as predicted by theorem 15.3.3; see figure 15.4.

The foregoing results were obtained by positing nonselective harvesting. Usually, because of the nonintentional harvesting of the other resources, nonselective harvesting is considered potentially more dangerous for the survival of them. In Semmler and Sieveking (1992) additional simulations are reported pertaining to selective harvesting. The results were that overall for the predator-prey system selective harvesting appeared to be more conservative than nonselective harvesting (whereas for the system of competing species we obtained the reverse result).

In sum, contingent on the elasticity of demand we find that for the open access case both the predator-prey system as well as the system of competing species admit limit cycles. These also occur for a sole owner regime (when the discount rate tends to infinity or is very large). Regarding the conservation of resources, we show that open access exploitation is generally inferior to the sole owner regime—as demonstrated for

[18] Note that for the monopoly case, with discount rate tending to infinity, already $\alpha = 1$ generates limit cycles that will be preserved for large discount rates. Thus, only in the case $\alpha = 1$ does a monopolist appear to be more conservative than the open access regime.

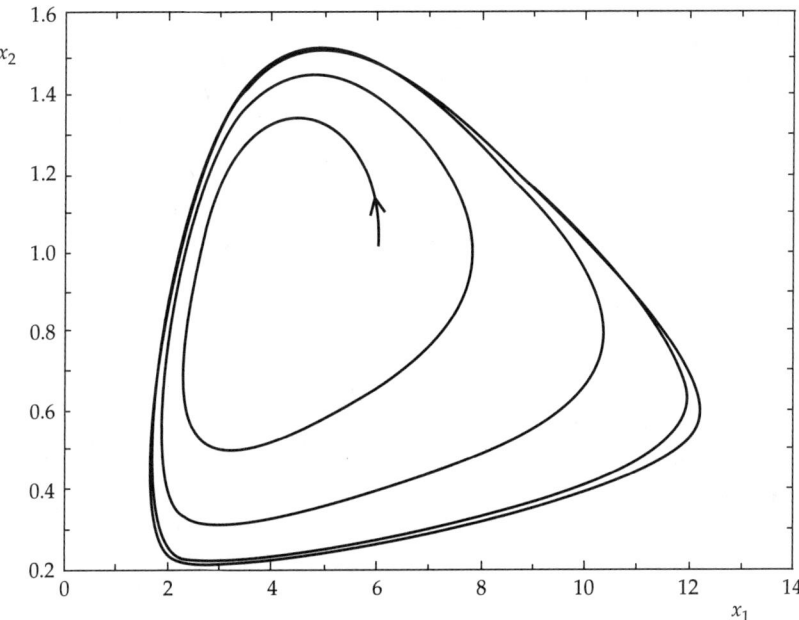

Figure 15.4 System of Competing Species, Open Access, $\alpha = 2$, Initial Values (8,1).

the predator-prey system—and nonselective harvesting appears to not necessarily be more destructive than selective harvesting.

15.4 THE MONOPOLY: INFINITE HORIZON OPTIMIZATION

Let us turn more specifically to the sole owner and the problem of the infinite horizon optimization. Here we expect, according to theorem 15.2.2, trajectories exhibiting limit cycles in the monopoly case occurring for $\delta \to \infty$ to be replicated,[19] at least for large discount rates. We study a sampling of trajectories and analyze model versions where all parameters are held fixed except the discount rate under which the optimization, as described by (15.9), takes place. According to the one species theory, a lower discount rate is less critical for the survival of a species (see Clark 1990, chap. 2.5). For two or more species it may be just the opposite, as the following simulations demonstrate. In fact, the limit $\delta \to 0$ may result in extinction of the species. Results for a varying discount rate are explored with respect to the systems of predator-prey

---

[19] Limit cycles in the monopoly case for the discount rate tending to infinity are reported in the previous section. Those simulations were employed as benchmark cases for studying the trajectories for finite discount rates.

and competitive interactions. We restrict our study to nonselective harvesting. A dynamic programming algorithm as described in Semmler and Sieveking (1992) is employed in computing the optimal trajectories.

Let us start again with the predator-prey system. Figure 15.5 depicts the monopoly case for $\delta = 5$.

As one can observe in the figure 15.5, the zero time-horizon optimization trajectory for the monopoly case, $\alpha = 1$ (which should be the same as in figure 15.2 for the open access, $\alpha = 2$), is replicated for infinite time horizon optimization with a large discount rate.

Further simulations with our dynamic programming algorithm were undertaken for the monopoly case in the predator-prey system with diminished discount rates. The results, for discount rates $\delta = 1, 5$ and $\delta = 0.05$ are shown in figures 15.6 and 15.7.

The limit cycles, depicting the optimal trajectories, appear to increase with smaller discount rates. If δ tends to be small, for example $\delta = 0.05$, one of the species tends to be driven to the axis.

Next we consider the system of competing species for a finite δ, again for the monopoly case only. Note that now the corresponding limit cycle for the monopoly with $\delta \to \infty, \alpha = 1$, is figure 15.4 (open access, $\alpha = 2$).

Figure 15.8 again exhibits a limit cycle for $\delta = 30$, resembling the corresponding figure 15.4.

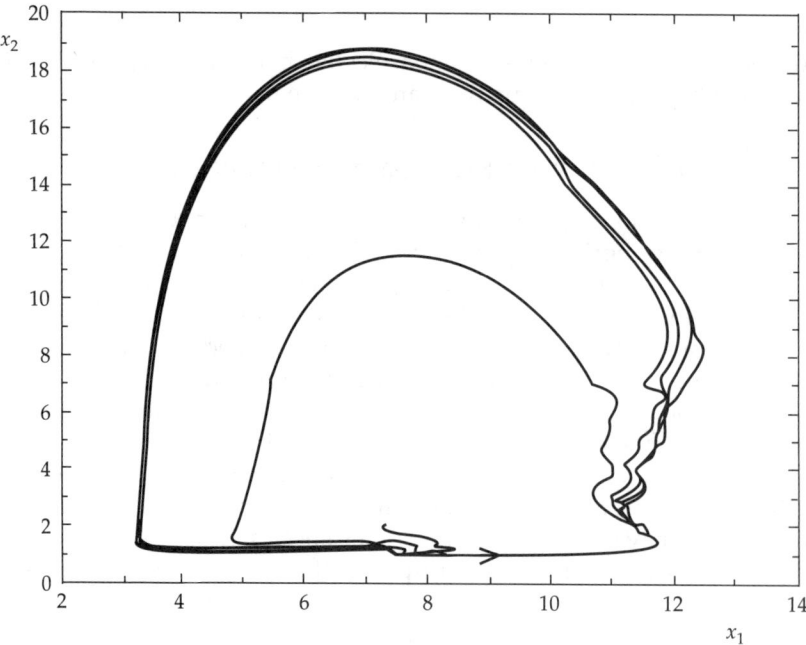

Figure 15.5 Predator-Prey, Monopoly, $\alpha = 1(\delta = 5)$, Initial Values (7,2).

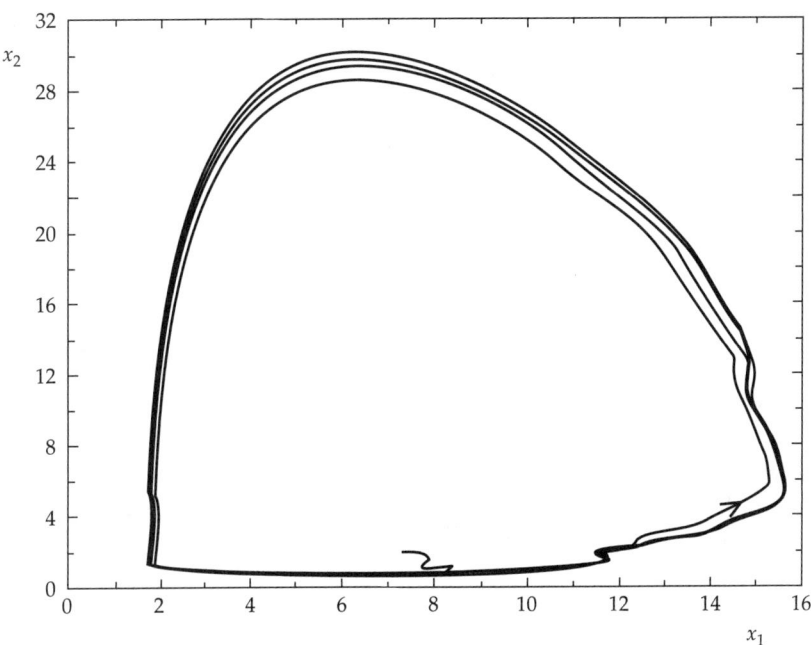

Figure 15.6 Predator-Prey, Monopoly, $\alpha = 1(\delta = 1.5)$, Initial Values (7,2).

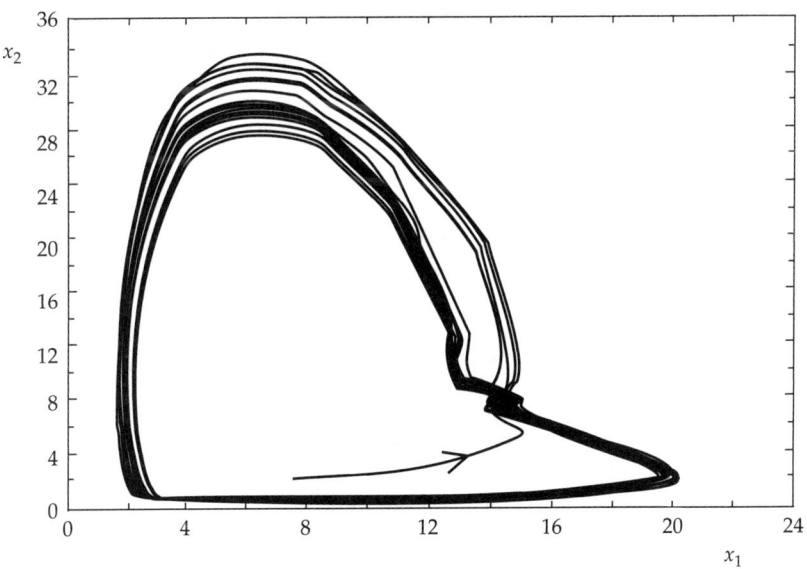

Figure 15.7 Predator-Prey, Monopoly, $\alpha = 1(\delta = 0,05)$, Initial Values (7,2).

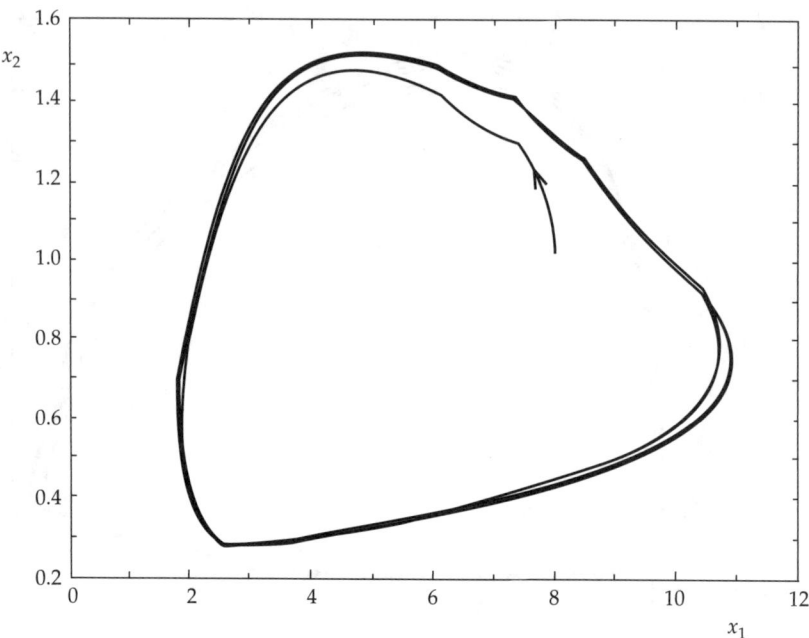

Figure 15.8 System of Competing Species, Monopoly, $\alpha = 1(\delta = 30)$, Initial Values (7,1).

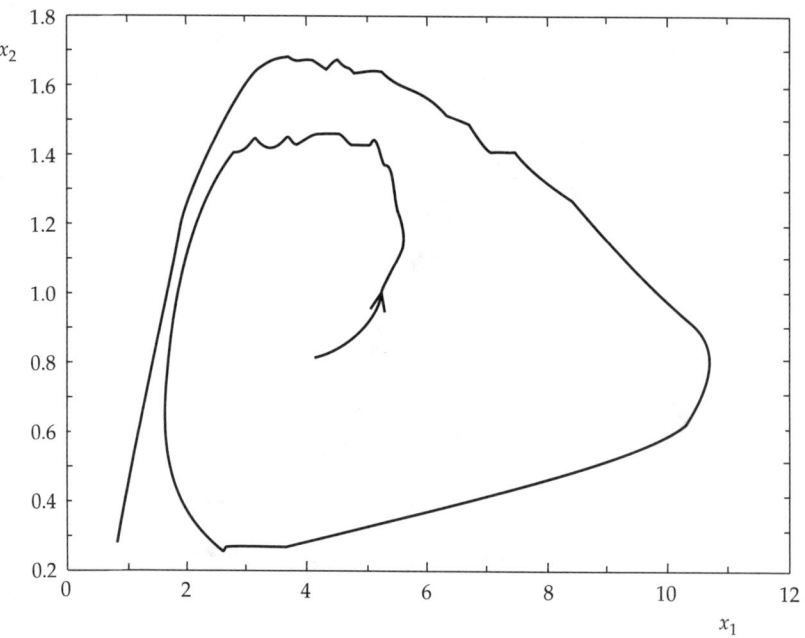

Figure 15.9 System of Competing Species, Monopoly, $\alpha = 1(\delta = 10)$, Initial Values (4,8).

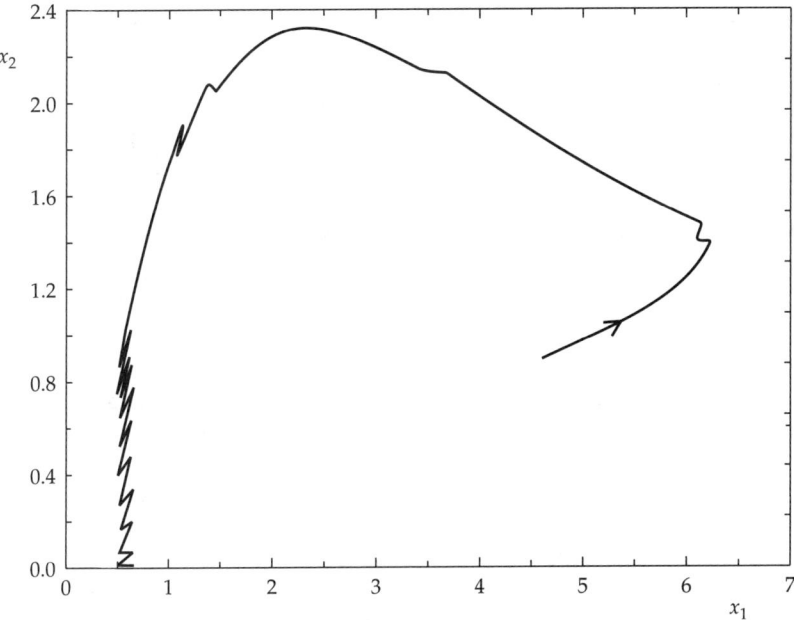

Figure 15.10 System of Competing Species, Monopoly, $\alpha = 1(\delta = 2)$, Initial Values $(5, 0.8)$.

However, as exhibited in figure 15.9 for a discount rate $\delta = 10$, the cycles tend to be more unstable, with one of the species driven to extinction.[20]

For a discount rate of $\delta = 2$, as figure 15.10 shows, one species appears to be completely driven to extinction.

As it turns out, in these examples, the discount rates do not have to be too large to replicate the limit cycles (in fact, in the predator-prey example, figure 15.7, it is rather small). Moreover, as shown, in the context of multispecies harvesting, lower discount rates do not necessarily give rise to a better resource conservation. Lower discount rates, as demonstrated for the systems of predator-prey and competing species may admit more unstable trajectories (and drive species to extinction).

15.5 CONCLUDING REMARKS

This chapter has studied the optimal exploitation of ecologically interrelated resources and demonstrates that propositions derived from

---

[20] From figure 15.10 it is visible that there are points of extinction on the x_1 and x_2 axes. The fact that for certain discount rates the trajectories converge toward one of the axes indicates that there are bifurcation values for δ at which a new dynamic is generated. Several initial conditions were tried for simulations such as depicted in figure 15.9. None of them continued to generate a limit cycle.

models of harvesting of a single isolated resource only do not carry over to the case of ecologically interacting resources.[21] As shown, the useful analytical distinction between zero time horizon and infinite time horizon optimization helps study the dynamics of interrelated resources and their final fate. The zero time-horizon optimization gives predictions for the behavior of the trajectories in infinite time-horizon optimization for large discount rates. How large the discount rate has to be to reproduce the results from the zero optimization problem is explored by computer simulations. Those simulations undertaken by the aid of a dynamic programming algorithm can indeed track the predicted trajectories well. The amplitude and frequency of persistent cycles (limit cycles)—if they exist—as well as extinction of resources can be investigated. Computer simulations are thus indispensable to explore, for example, for various discount rates the fate of interacting resources when harvested under the different regime studied.

Furthermore, our study revealed several substantive results of interest for resource economists and resource management. Our analytical (but in particular our numerical) study of optimal exploitation of interacting resources showed that (1) monopoly ownership is not likely to conserve resources better than open access relations; (2) selective (in contrast to nonselective), harvesting will not always lead to superior results regarding the conservation of resources; (3) lower discount rates may, contrary to what one might expect from a single resource, give rise to greater instability of the trajectories and possibly lead to depletion of resources; and (4) the dynamics of resources depend on the elasticity of demand when the resource is sold on the market. We may tentatively summarize that the results of this study point to the conclusion that management policies for a single resource are not necessarily prudent in many circumstances for interacting resources.

Finally we note that when environmental resources are viewed as environmental assets, our method of using dynamic programming can also compute the value of the value function at each point of the two-dimensional state space (if we have two interacting resources) and thus the asset value of the resources. This is basically the discounted future payoff along the paths that are computed. Such a computation is undertaken in Grüne and Semmler (2004) where the value function, in addition to the dynamics of the state equations, are computed and shown how the asset value behaves depending on the dynamics of the state variables. The latter, as our examples in section 15.4 showed, can follow a rather intricate dynamics, and one expects the same for asset value of

---

[21] Note that optimization models for economic growth and resource extraction also have mostly considered only one (homogenous) resource; see the seminal work by Hotelling (1931), Dasgupta and Heal (1974), and Koopmans (1985). It would be of interest to recast those types of models for the case when the extraction of interrelated resources is admitted.

the environmental resources. Yet because dynamic programming solves for the value function, it can also answer crucial questions on the pricing of environmental assets.[22]

Appendix

A.1 OPEN ACCESS OPTIMAL EFFORT (THEOREM 15.3.1)

Consider the function $v_\alpha(x)$ as proposed in (15.15) with

$$p_\alpha(y) = (y_0 + y)^{-\alpha}; \ \alpha > 0, \quad y = xv(x)$$

for $G(x, v) = p_\alpha/xv(x)) - (c/x) = 0$; then

$$v_\alpha(x) = x^{-1}[(cx^{-1})^{-1/\alpha} - y_0]$$
$$= x^{-1}[c^{-1/\alpha}x^{1/\alpha} - y_0]$$
$$= c^{-1/\alpha}x^{1/\alpha-1} - y_0 x^{-1},$$
$$v_\alpha = 0 \text{ for } 0 \le x \le y_0^\alpha c,$$

and

$$v_\alpha = c^{-1/\alpha}x^{1/\alpha-1} - y_0 x^{-1} \quad \text{for } y_0^\alpha c < x;$$
$$v'_\alpha(x) = \frac{1-\alpha}{\alpha}c^{-1/\alpha}x^{1/\alpha-2} + y_0 x^{-2}$$

and

$$v'_\alpha > 0 \Leftrightarrow y_0 \ge \frac{\alpha - 1}{\alpha}c^{-1/\alpha}x^{1/\alpha}.$$

Therefore, $v'_\alpha(x) > 0 \Leftrightarrow 0 < \alpha \le 1$ and decreasing for $\alpha > 1$ because in the latter case $v'_\alpha < 0 \Leftrightarrow x > y_0^\alpha c \left(\frac{\alpha}{\alpha-1}\right)^\alpha$.

A.2 THE MONOPOLIST'S OPTIMAL EFFORT $\delta \to \infty$

With $\delta \to \infty$ for the monopolist's optimal effort, we have

a. There is a unique solution $u_\alpha(x)$ to 15.11 that is a smooth function of $x \ge c y_0^\alpha$; $u_\alpha(x) = 0$ for $0 \le x \le c y_0^\alpha$.
b. The optimal supply $y_\alpha(x) = x u_\alpha(x)$ is increasing in x and satisfies $\lim_{x\to\infty} y_\alpha(x) = y_0/\alpha - 1$ for $\alpha > 1$ and $\lim_{x\to\infty} y - \alpha(x) = +\infty$ else.

---

[22] For explicit examples of computing the asset value through dynamic programming, see Grüne and Semmler (2004) and Becker et al. (2007).

c. For $\alpha = 1$,

$$u_1(x) = c^{-1/2}\gamma^{1/2}x^{-1/2} - \gamma_0 x^{-1};$$

$$\max u_1 = \frac{1}{4c}, u_1'(x) < 0 \quad \text{for} \quad x > 4\gamma_0 c.$$

d. If $\alpha > 1, u_\alpha(x)$ is increasing-decreasing with $\lim_{x \to +\infty} u_\alpha = 0$; if $4/5 < \alpha < 1$, then $u_\alpha(x)$ is increasing-decreasing-increasing with $\lim_{x \to +\infty} u_\alpha(x) = +\infty$; if $0 < \alpha < 4/5$, then $u_\alpha(x)$ is increasing with $\lim_{x \to +\infty} u_\alpha(x) = +\infty$.

The case $\alpha = 1$ then admits a limit cycle. Note that for $\alpha = 1$ the effort $u_\alpha(x)$, in the case of the monopolist, is the same for the open access case with $\alpha = 2$. For proofs of these propositions, see Semmler and Sieveking (1994b).

A.3 LIMIT CYCLE FOR THE PREDATOR-PREY SYSTEM (THEOREM 15.3.2)

Consider the function $H = a_2\bar{x}_2\log x_2 + b_1\bar{x}_1\log x_1 - b_1 x_1 - a_2 x_2$. Note that $H_{x_1}x_1(a_2\bar{x}_2 - a_2 x_2) + H_{x_2}x_2(-b_1\bar{x}_1 + b_1 x_1) = 0$. Note also that together with $v_1(x_1)$ and $v_2(x_2)$, the functions

$$\bar{v}_1 = v_1(x_1) + a_1 x_1, \quad \bar{v}_2 = v_2(x_2) + b_2(x_2)$$

are increasing at $\bar{x}_1, \bar{x}_2$. Therefore,

$$\frac{d}{dt}H(x(t)) = H_{x_1}x_1(a_0 - a_1 x_1 - a_2 x_2 - v_1(x_1)) +$$
$$+ H_{x_2}x_2(-b_0 + b_1 x_1 - b_2 x_2 - v_2(x_2))$$
$$= H_{x_1}x_1(a_1\bar{x}_1 + v_1(\bar{x}_1) - a_1 x_1 - v_1(x_1))$$
$$+ H_{x_2}x_2 b_2\bar{x}_2 + v_2(\bar{x}_2) - b_2 x_2 - v_2(x_2))$$
$$= b_1(\bar{x}_1 - x_1)(\bar{v}_1(\bar{x}_1) - \bar{v}_1(\bar{x}_1))$$
$$+ a_2(\bar{x}_2 - x_2)(\bar{v}_2(\bar{x}_2) - \bar{v}_2(\bar{x}_2)) > 0.$$

Hence H is a Liapunov function on $(0, +\infty)^2$ and assertion a follows. Also, b holds. We use the same Liapunov function as before. The hypothesis that v_i is locally decreasing at $\bar{x}_i$ implies that any nonstationary trajectory near $(\bar{x}_1, \bar{x}_2)$ spirals away from $(\bar{x}_1, \bar{x}_2)$. It must then tend to a limit cycle provided the trajectory is bounded, which it is because of $a_1 + b_2 > 0$.

A.4 EQUILIBRIA FOR THE SYSTEM OF COMPETING SPECIES

Let $(x_1, x_2) = e$ be a stationary state with positive coordinates. The Jacobian at such a state is

$$\begin{bmatrix} -x_1(a_1 + v_1'(x_1)) & -x_1 a_2 \\ -x_2(b_1 + k_1 v_1'(x_1)) & -x_2 b_2 \end{bmatrix}.$$

The roots if the corresponding characteristic polynomial are

$$\lambda_\pm = -\frac{1}{2}[x_1(a_1 + v_1'(x_1)) + x_2 b_2] \pm \sqrt{d},$$

where $d = (1/4)[x_1(a_1 + v_1'(x_1)) - x_2 b_2]^2 + x_1 x_2 a_2(b_1 + k_1 v_1'(x_1))$.

Thus, if $v_1'(x_1) > 0$, x will be stable or a saddle point. In fact, x will be a saddle point if and only if

$$(a_2 k_1 - b_2)v_1' > a_1 b_2 - a_2 b_1.$$

To simplify, we assume $a_1 = b_2 = 0$.

If limit cycles turn out, small $a_1 > 0, b_2 > 0$ may be reintroduced without destroying them, see Sieveking (1990).

Let $\phi(x_1) = b_0 - b_1 x_1 - k_1 v_1(x_1)$.

If ϕ has two or more zeros, then it has at least three zeros on $[0, +\infty)$ because since we assume $v_1 \geq 0$. At the first zero, say w_1, $b_1 + k_1 v_1'(w_1) > 0$ and the corresponding stationary state $w = (w_1, w_2)$ is a saddle. At the second, say, $y = (y_1, y_2)$, $b_1 + k_1 v_1'(y_1) < 0$. Therefore, y is repelling. A third one, $z = (z_1, z_2)$, will again be a saddle. In the case of three stationary states with positive coordinates, different phase portraits may be classified according to the relative position of the separatrices emanating from the saddles.

16

Regulation of Resources

16.1 INTRODUCTION

The work of Clark (1971, 1990) has demonstrated that government resource management is in particularly needed to avoid the depletion of renewable resources. Also the last chapter showed that resource management appears to be indispensible regardless of whether resources are common property or owned privately.[1] However, the regulatory means by which resources are managed are controversial. For common property resources, a large number of instruments have been proposed for pursuing a policy of resource conservation.[2] Among them are quota and licensing schemes, issuing and trading permits, levying taxes, formal entry restrictions for new firms, and subsidization of the exit of firms.

In this chapter we exemplify the problem of resource regulation by studying the tax rate as regulatory instrument. Tax rates and other regulatory instruments to reduce pollution were studied in chapter 10. For the use of the tax rates in ecological management problems, see Grüne et al. (2005). In this chapter we study how tax rate on operating firms can prevent renewable resources from depletion. The model underlying our considerations here originates in a study by Clark (1990, pp. 118–122). In the present chapter some of Clark's conjectures are explored. Moreover, an extended version of the original Clark model is presented, and a dynamic programming algorithm is applied that replicates (by means of computer simulations) the model as well as an extended version.

Three features of the modeling approach to study regulatory problems are worth mentioning. First, the regulated resource is assumed to be renewable. It is extracted through an effort and a constant cost per unit of effort spent. Firms that extract the resource behave optimally. They maximize the present value of a return from their action in the context of an infinite horizon optimization model. Also, for reasons of simplicity, as commonly used in models of resource exploitation, we

---

[1] A careful reading of Hotelling (1931) will reveal that even for exhaustible resources some "state interference" (Hotelling 1931, p.143) is needed, for example, in the form of restriction of resource exploitation or taxation.

[2] For an extensive discussion, however, in a static framework, see Dasgupta and Heal (1979), ch. 3.

assume the output price to be fixed.[3] Because there is only one resource, complicated interactions between different types of resources while one is extracted are disregarded. As shown in the last chapter, the exploitation of interacting resources usually generates more intricate dynamics. An optimization model for this case but without regulatory policies is presented in Semmler and Sieveking (1994b) and was also studied in the previous chapter.

Second, the effort spent in extracting the resource is represented by a variable capacity utilization rate that is endogenously determined. Effort is thus not a control variable in the optimization model but a state variable. This approach originates in Clark (1990, pp. 118–122) and Clark et al. (1979) where the effort cannot be reduced instantaneously by reducing the rented capital stock. With irreversible investment decisions, the capital stock might be reducible only slowly as argued in Clark et al. (1979), who relate their model back to a contribution by Arrow and Kurz (1970) on capital accumulation with irreversible investment.[4] In our work, too, the capacity can be adjusted only slowly and therefore cannot be a choice variable for the firms extracting the resource. This entails that firms also continue to undertake the operation, though they temporarily face negative profits.[5] In a more complete version of our model then, we keep track of the budget constraint of the firms so that in addition to the stock of the resource and the utilization rate, the evolution of the debt of firms also enters the dynamics of the state variables.

A third feature of our model springs from the regulatory policy proposed in this chapter. To levy a fixed tax on the extraction of a resource is surely not optimal and might not necessarily conserve the resource. What we discuss along the lines of Clark (1990, pp. 118–122) is an optimal tax. In the limit case, when the discount rate tends to infinity, the optimal tax turns out to be of the bang-bang type: a simple feedback control of the tax rate with the tax rate high when the resource tends to be depleted and low when the stock of the resource is large. This bang-bang policy (frequently employed as a rule of thumb) is shown to be optimal. The varying tax rate also turns out to be optimal for the case of a finite discount rate.[6] A cyclical tax rate is thus the optimal regulatory

---

[3] A variable price determined by an inverse demand function can easily be included in such an approach. See Semmler and Sieveking (1994b).

[4] For an excellent survey on the theory and empirical evidence of irreversible investment, see Pindyck (1991).

[5] The fact that firms can be taxed more easily if investment is irreversible was already discussed in Kalecki (1938) who derives irreversible investment from the "time to build" or the gestation period of investment goods.

[6] Recently a number of papers have been published that study optimal taxes in the context of two dimensional optimal growth models as originating in Lucas (1988) and Romer (1990). Here it is shown that a consumption or lump-sum tax are nondistortionary, whereas the optimal income or profit taxes are zero. This means that a tax rate on income or

policy with a heavy tax rate applied when the stock of the resource is low and a low tax rate when the stock of the resource is high. Indeed, with certain assumptions on the "production function" governing the stock of the resource, limit cycles can be proved to exist for the stock of the resource, the utilization rate, and the optimal tax policy.

Though we focus here only on one regulatory instrument to regulate resources, the result may lend itself to wider implications. A variety of public interventions, including aspects of quota policies, subsidies, entry restrictions, property right restrictions, and other regulatory means, such as issuing and trading of permits, may be studied in a similar manner. The results might potentially be interpreted as bearing on other aspects of governmental actions.

Section 16.2 lays out a three-dimensional model of resource exploitation that originates in Clark's contribution. Clark's original model does not introduce a budget constraint of firms and gives no analytical results regarding persistent cycles. For the two-dimensional variant, we analytically demonstrate that the optimal tax is cyclical. The analytical study of the partial system provides us with predictions of how the trajectories of the full system will behave (see appendix). Section 16.3 demonstrates, by employing a dynamic programming algorithm, that the cyclical trajectories remain preserved even if a dynamic equation for the evolution of debt is added to the dynamics. Section 16.4 provides some concluding remarks.

16.2 THE MODEL

Our intertemporal model of the extraction of resources originates in Clark (1990, p. 118). There is a present value to be maximized. The present value is the discounted stream of revenues of the firms, for example, in an industry extracting a renewable resource. There are three state equations: one represents the law of motion for the exploitable resource (in Clark, for example, stock of fish); the second depicts the capacity utilization or effort E by which the resource is exploited.[7] Because we allow firms to borrow from the capital market to close the budget gap,

---

a profit tax are inversely related to the growth rate of the economy. In the papers by Chamley (1986), Judd (1987), Rebelo (1991), King and Rebelo (1990), and Lucas (1990) the effects of taxes—particularly capital tax—on growth are studied. Frequently, these models, however, provide solely a steady-state analysis. With government expenditures viewed as pure consumption, it is then derived that capital tax should be zero at the steady state. However, as in Jones et al. (1993) and Greiner et al. (2005, ch. 6) when government expenditure is treated as productive input, then the optimal tax on capital also can be nonzero at the steady state.

[7] A formulation of a model where the change of effort (the number of firms or amount of capital) is a sign preserving function of the excess profit earned in resource extraction was first set up by Smith (1968) and later development further by Clark (1990) and Berck (1979). Some of the models in this tradition assume additionally that for each competitive

there is a third state equation that characterizes the evolution of debt of
the firms. The tax rate is the control variable (the optimality of which
Clark has conjectured in a two-dimensional system.) The assumption
of a certain shape of the growth function F (see appendix) allows the
description of the law of motion of the resource stock. The model reads
as follows:

$$(*) \begin{cases} V_{\max} \int_0^\infty e^{-\delta t} G(x_t.E_t, D_t) dt \\ \text{with } G(x_t, E_t, D_t) = (px_t - c)E_t - \phi(iD_t) \\ \text{subject to} \\ \dot{E} = kE_t{}'[(p - \tau)x_t - c]. & (16.1) \\ \dot{x} = x_t[F(x_t) - E_t], & (16.2) \\ \dot{D} = iD_t - [(p - \tau)x_t - c]E_t, \\ \tau_{min} \le \tau_{max} \quad x(0) = x_0, \quad E(0) = E_0, & (16.3) \end{cases}$$

where p is the price (taken as fixed), c is a constant extraction cost per
unit of effort spent (c includes a rental price of capital), τ is the tax
rate, $k > 0$ is a constant, and $p > \tau_{max}$. Similarly as in Clark (1990,
p. 118) the objective is to maximize the (net) rent flow $G(x_1, E_1, D_1)$ of
the extractive industry, which is then divided between the extractive
industry, which receives the portion $[(p - \tau)x_1 - c]E_1$, and the taxing
authority, which receives the remainder $\tau x_1 E_1$. Note that the control
variable τ does not appear under the integral. D is the debt of firms and
i is the interest rate that is assumed to be different from the discount
rate δ. Debt enters in the return function $G(x_1, E_1, D_1)$ that affects the
net present value of the industry.[8] For reasons of simplicity, we assume
$\phi(iD_1) = iD_1$. Equation (16.1) represents the evolution of the effort spent
E and can be read as the capacity of the industry. Because x denotes the
stock of resource, equation (16.2) depicts the change of the stock with
$F(\cdot)$ describing its growth function and E denotes the effort by which
the stock is reduced.

Equation (16.3) denotes the change of debt of the industry, the first
expression being the debt service and the second the rent flow of the
resource industry. Thus, firms borrow when the debt service is greater
than the rent flow, otherwise, when they accumulate financial assets and
receive an interest payment iD_1. Moreover, the tax τ is constrained to be
between a feasible maximum and minimum level. This model will be
studied further in section 16.3.

The partial model put forward by Clark (1990, pp. 118–122) includes
only two state equations: one for the exploitable resource and the other

---

firm the price is equated instantly to marginal cost, and then with fixed cost present, the
excess of price over average cost will encourage entry.
[8] For details of such an approach see Asada and Semmler (1991).

for capacity or effort E, with the tax rate τ as the control variable. The model reads as follows:

$$
(**) \begin{cases}
V_{\max} \int_0^\infty e^{-\delta t} G(x_t.E_t, D_t)\,dt \\
\text{with } G(x_t, E_t, D_t) = (px_t - c)E_t - \phi(iD_t) \\
\text{subject to} \\
\dot{E} = kE_t[(p - \tau)x_t - c]. & (16.4) \\
\dot{x} = x_t[F(x_t) - E_t], & (16.5) \\
\tau_{min} \le \tau_{max} \quad x(0) = x_0, \quad E(0) = E_0.
\end{cases}
$$

The notations are the same as in the previous section.

According to Sieveking and Semmler (1994b) there exists a limit system of $(**)$ that is approached for the discount rate $\delta = \infty$. For such a limit system, called system $(***)$ in the appendix, the trajectories are tractable analytically. The analysis in the appendix establishes that for systems $(***)$ and $(**)$ there exist periodic solutions—limit cycles—for the tax rate and the two state variables E and x. Two more specific results are established in Semmler (1992) by way of computer simulations. First, the size of limit cycles depends on the discount rate: with a lower discount rate, the periodic solution shrinks to an attractor. Second, the value function exhibits, in certain regions, negative values.[9] The simulations are undertaken by a dynamic programming algorithm as presented in the appendix of Sieveking and Semmler (1994b).[10]

16.3 NUMERICAL RESULTS FOR THE PARTIAL AND COMPLETE MODEL

It is of great interest to study the optimal trajectories for the complete system $(*)$, the three-dimensional system, where firms face a properly formulated budget constraint and where the debt feeds back to the value function as in Asada and Semmler (1995). By way of the dynamic programming algorithm, we are able to show that the results of system $(**)$ carry over to system $(*)$. Figures 16.1 and 16.2 show a sampling of trajectories.

---

[9] In technical terms, the phenomenon that firms continue to stay in an industry even as their present value becomes negative, has been called hysteresis. Sunk costs to open and close an operation can explain the "hysteresis often observed in extractive resource industries: During periods of low prices, managers often continue to operate unprofitable mines that had been opened when prices were high; at other times managers fail to reopen seemingly profitable ones that had been closed when prices were low," Pindyck (1991), p. 1134, see also Dixit (1989).

[10] The working paper with the description of the dynamic programming algorithm is available on request.

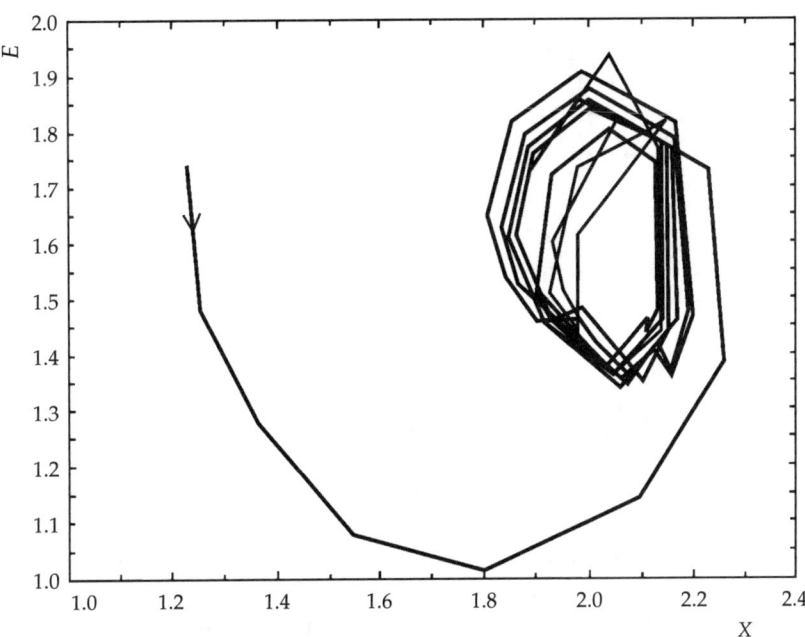

Figure 16.1 Limit Cycle for (x, E) of System $(*)[\delta = 1,\ D(0) = 0.1]$.

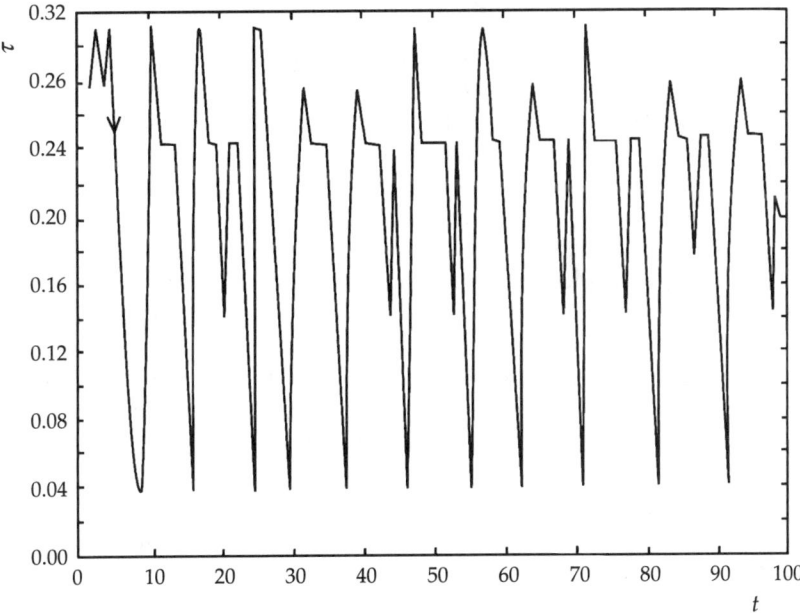

Figure 16.2 Optimal Tax Rate τ from System $(*)[\delta = 1,\ D(0) = 0.1]$.

Figure 16.1 depicts the trajectories for x and E for the optimally controlled system ($*$) when $\delta = 1^{11}$ and the initial condition $D(0) = 0.1$.

As can be observed also for the system ($*$) in the plane, there is a limit cycle for (x, E). The value function as obtained from our dynamic programming algorithm for the three-dimensional case ($*$) has roughly the same shape as reported for the system ($**$) in Semmler (1994). It also showed regions of negative values. Because of space limitation, we forgo a graph of it.

Next, we want to demonstrate the time path of the optimal tax rate resulting from system ($*$). As can be seen in figure 16.2, the optimal tax rate is cyclical.

Thus, as depicted in figures 16.1–16.2, the limit cycle that result from system ($**$) remains preserved for system ($*$), that is, even the evolution of debt—with a feedback to the value function—is added. In the numerical study for system ($*$) we could, however, observe a slightly unstable trajectory of the debt, D_t. This requires to study the critical level of debt.[12] Empirically, one would not expect the debt to increase without bounds. Firms in the industry would lose credit-worthiness. Thus, realistically, one should assume some bounds for an unstable trajectory D_t. These bounds are likely to feed back to the dynamics of system ($*$) and prevent a further explosion of D_t. This is likely to keep the dynamics (a limit cycle for x, E, and τ) inside those bounds.[13]

16.4 CONCLUDING REMARKS

In the context of an intertemporal optimization model with irreversible investment, we show that optimal policies for protecting resources from extinction might be cyclical. This is demonstrated to hold for an optimal tax rate in the context of a static optimization problem (zero horizon optimization problem with discount rate tending to infinity) as well as an infinite horizon problem (finite discount rate). A periodic tax rate is the optimal regulatory policy with a heavy tax when the resource stock is low, and a low tax rate is applied when the stock of resource is large. Indeed, with certain assumptions on the production function governing the growth of the resource, the optimal solution paths are periodic. This holds for the capacity utilization rate, the stock of the resource, and the tax rate. These results appear to hold true for the original two-dimensional Clark model as well as for a three-dimensional

---

[11] Cyclical trajectories are also obtained for smaller discount rates; see Semmler (1994).

[12] Such a critical level of debt, although in a different model, is further explored in Semmler and Sieveking (1994b).

[13] It is easy to reformulate system ($*$) in a way that it turns into different dynamics when the debt becomes too high but where for a level of debt below an upper bound the old dynamics ($*$) remain intact. This way the dynamics would only change at outer boundaries, keeping a limit cycle inside those boundaries.

extension of it that includes a budget constraint of firms in the extractive industry.

Technically, the problem under consideration is of broader interest. Two methods are employed that admit a study of the out-of-steady-state dynamics of an intertemporal optimization problem. First, the infinite horizon optimization problem is reduced to a zero horizon problem where the control variable can be computed explicitly in feedback form and the system's dynamics can be studied analytically; see the appendix. Second, a dynamic programming algorithm permits computation for our chosen parameter constellations of the optimal trajectories for the infinite horizon problem with a finite discount rate. The dynamics from the first system (zero horizon problem, system (* * *)) carry over to the second system (infinite horizon, finite discount rate, system (**)). In addition, by way of computer simulations, it is also shown that the dynamics reappear for the three-dimensional system where the evolution of financial structure of firms is modeled in a third differential equation, system (*). Another technical aspect is also of interest. In the model, the control variable does not appear in the return function, but rather in the state equations. An appropriate transformation, however, can be shown to lead back to a common optimization problem with the control variable included in the return function.

Last, we want to note that there are, of course, other regulatory instruments for controlling the overuse of renewable resources. In resource economics, in addition to taxes, the effects of property rights, quota, and license schemes issuing and trading of permits as well as instruments limiting entry into extractive industries have been discussed extensively.[14] Most of the studies on the regulation of extractive industries, however, do not consider the effects of the regulatory instruments in the context of a dynamic decision model, and it is therefore not clear whether the advantages and disadvantages of regulatory policies discussed in the static model would carry over to dynamic models. For example, employing quotas or entry restrictions to an extractive industry, in the context of our model, will affect industry effort E and this introduce an upper bound for the state variable E. Without going into a deeper analysis of this problem, one can conjecture that the cyclical solutions resulting from a control by taxation will most likely be altered, depending, however, on the specification of the kind of quota scheme employed. Preliminary work along this line, with one state variable only, has been undertaken by Clark (1990, ch. 8). Also in the context of our model, this would be a worthwhile question to pursue further in a future study.

---

[14] For an elaborate discussion of advantages and disadvantages of different regulatory means, see Dasgupta and Heal (1979, ch. 3) and Clark (1990, ch. 8).

Another important regulatory instrument is currently discussed in the context of how to control carbon emission that is now proved to give rise to global warming. In part II of the book, we have discussed tax rates and abatement policy instruments. Recently, to fulfill the Kyoto agreement on the reduction of carbon emission until 2012, the trading of carbon emission has been introduced as regulatory instrument in Europe. National governments in the EU are required to agree (since 2005) to reduce carbon emission in certain industries, for example, steel, chemical, and electricity production. Some success seems to be visible, but whether this instrument will effectively reduce carbon emission in the long run remains to be seen.[15]

Appendix

The limit system of $(* * *)$, with $\delta \to \infty$, is

$$(* * *) \begin{cases} \max G(x, E, \tau) \\ \text{subject to} \\ \dot{E} = kE_t[(p - \tau)x_t - c]. & (16.6) \\ \dot{x} = x_t[F(x_t) - E_t], & (16.7) \\ \tau_{min} \leq \tau_{max} \quad x(0) = x_0, \quad E(0) = E_0. \end{cases}$$

Thus, with $\delta \to \infty$, a static optimization problem arises.[16] To obtain $(* * *)$, we employ partial integration after writing $y = (x, E)$. Express the integral in $(**)$ as

$$\int_0^x e^{-\delta t} G(y_t) dt = -1/\delta G(y_0) + 1/\delta \int_0^x e^{-\delta t} G'(y_t) \dot{y} dt.$$

Thus $G(y_t) = (px_t - c)E_t$ is replaced by the expression

$$G(x, E, \tau) 0 px[F(x) - E]e + (px - c)kE[(p - \tau)x - c]. \quad (16.8)$$

When $\delta \to \infty$ and $(* * *)$ is obtained, (16.8) is to be maximized. The derivative of (16.8) with respect to τ yields

$$-(px - c)kE\tau x \quad (16.9)$$

(16.8) is maximized if

$$\tau = \begin{cases} \tau'_{min} & \text{for } px \geq c, \\ \tau'_{max} & \text{for } px \leq c. \end{cases}$$

---

[15] In particular, a cap on carbon emissions in industries leaves open the questions of carbon emission in energy production, transport, households, and other areas of economic and social life.

[16] A theorem explaining this procedure is presented in Sieveking and Semmler (1994b).

For the optimization with zero time horizon, as represented by (***), a bang-bang control arises.

For (* * *) the following three equilibria exist:

$$(E0) : (0,0). \quad (E1) : (x_1, F(x_1)), \quad (E2) : (x_2, 0)$$

with $x_1 = c/(p - \tau_{min})$ and $x_1 < x_2$.

The characteristic polynomial of the Jacobian, for example, at the equilibrium $(E1)$ is

$$\chi_{E1}(\lambda) = \lambda^2 - x_1 F'(x_1)\lambda + ckF(x_1),$$

with roots $1/2x_1 F'(x_1) = \sqrt{1/4x_1^2[F'(x_1)^2 - ckF(x_1)]}$. Hence if $F'(x_1) > 0$, then $(E1)$ is a repeller. The characteristic polynomial at (2) is

$$\chi_{E2}(\lambda) = (x_2 F'(x_2) - \lambda)((p - \tau_{min})x_2 - c - \lambda),$$

with $\lambda_1(E2) = x_2 F'(x_2)$ and $\lambda_2(E2) = (p - \tau_{min})x_2 - c$.

A phase portrait in Semmler (1994) shows that there is a invariant compact set. All trajectories will enter this set. Moreover, if $f'(x_1) > 0$, all trajectories may not tend to one of the equilibria $(E0)$, $(E1)$, or $(E2)$ and must therefore tend to a limit cycle. (For limit cycles in related model, see Feichtinger et al. [1991]).

There are two interesting cases.

Case 1: $F(x) = rx(1 - xK^{-1})$. Then $x_2 = K$ and $F'(x_2) > 0 \leftrightarrow c < 1/2K$ $(p - \tau_{min})$. In this case there is a limit cycle for (* * *), hence, with sufficiently large δ also for (**).[17]

Case 2: $F(x) = r(1 - xK^{-1})$. Consider

$$H(x, E) = q(x - x_1 \ln x) + E - F(x_1)\ln E,$$

where $q = (p - \tau_{min})k$. H is a Liapunov function for (***). Then

$$H(x, E) = \frac{\partial H}{\partial x}\dot{x} + \frac{\partial H}{\partial E}\dot{E}$$

$$= q\left(1 - \frac{x_1}{x}\right)x(F(x) - E) + \left(1 - \frac{F(x_1)}{E}\right)kE[(p - \tau_{min})x - c]$$

$$= k(x - x_1)(F(x) - F(x_1)).$$

H is a Liapunov function provided that $(x - x_1)(f(x) - F(x_1)) < 0$ for $x \geq 0$.

Thus, in case 2, $(E1)$ is a global attractor for the limit system (* * *) on $(0, +\infty)$. In fact, asymptotic stability was conjectured by Clark (1990, p. 118) also for finite $\delta > 0$.

---

[17] The latter conclusion is shown to hold in Semmler and Sieveking (1994b).

17

Conclusion

Since the industrial take-off of many countries more then 150 years ago, the industrialization of countries in the North has increased their income per capita by roughly 2 percent a year. Since the end of the nineteenth century, the United States, for example, experienced an increase of per capita income by a factor of 10. The industrialization has spread in the twentieth century with many Asian countries taking off and catching up with the advanced countries. Yet the long-run process of economic growth and the rapidly spreading globalization in the past decades have also exhausted nonrenewable and renewable resources and led to the deterioration of the environment in some parts of the world. There is a rising concern of academics and politicians that the exhaustion of resources, the global change of the environment, and the climate change may have detrimental effects on economic activity and living conditions of future generations. To study those issues in a transparent and coherent framework, we designed small-scale models that have allowed us to study those issues and the implications for growth, environmental, and resource policies in a proper way.

In the first part of the book we analyzed in a small-scale model the interrelation between economic growth, the environment, and welfare of a country. We also introduced and discussed fiscal policy—income taxation, public investment, and abatement policies—and show how those can be used as instruments to improve the environment as well as economic growth. More specifically, we studied variants of models of endogenous growth with productive public spending and pollution, where we assume that pollution affects the utility of the household sector but not production possibilities directly. This assumption was made because it is obvious that environmental degradation affects growth if it has negative repercussions on production possibilities, whereas a growth effect is not so obvious if only utility is affected by pollution.

The analysis has demonstrated how a good policy can influence economic growth and welfare in the long run. We have seen that an increase in the pollution tax rate may raise the growth rate if sufficiently enough of the additional tax revenue is used for productive public spending, which stimulates the incentive to invest. This also means that a higher pollution tax rate, leading to a smaller increase in effective pollution, can reduce the balanced growth rate. Furthermore, we show that growth and welfare maximization may be different goals, particularly if the fiscal policy under consideration exerts a direct effect on the level of pollution. This result may be of particular relevance for developing countries and fast growing economies such as some Asian countries at the end of the twentieth century and the beginning of the twenty-first century. Concerning the dynamics of our model, we could demonstrate that indeterminate equilibrium paths may emerge, depending on the strength of the effect of pollution. Thus, pollution not only affects the growth rate in the long run, but may also be crucial as far as the transitional growth rate as well as regarding the question of to which growth rate the economy converges. Finally, we demonstrated that fiscal policy may lead to overshooting the variables, implying the transitional effects of economic policy differ from its long-run effects. Hence, a fiscal policy that raises the balanced growth rate may lead to temporarily smaller growth rates of economic variables.

We also analyzed a model economy where the environmental quality is either constant, improves, or deteriorates over time. The analysis of this model has demonstrated that sustained economic growth is compatible with a constant or improving environmental quality only if the production technology in use is not too polluting. It also turned out that an economy with a cleaner environment can only generate a higher balanced growth rate if the intertemporal elasticity of substitution of consumption is relatively high. This holds because with a high intertemporal elasticity of substitution, the marginal utility of consumption is higher in the future in those scenarios where the environmental quality is relatively clean. Therefore, the household is more willing to forgo consumption and shift it into the future in those scenarios implying a cleaner environment.

In the second part of the book, we studied the interaction of growth and global warming. This is also done within the framework of endogenous growth models. We presumed a simple endogenous growth model in which positive externalities of physical capital are the source of persistent growth. As for the modeling of the environment and the climate system, we integrated an energy balance model in the respective economic model and posited that deviations of the actual temperature from the preindustrial level imply damages by negatively affecting aggregate production. We analyzed both descriptive growth models as well as models with optimizing agents. An important result in this respect is that we could derive (which turned out to be quite robust) that economies with more polluting technologies should spend relatively more for abatement activities but nevertheless can emit more greenhouse gases compared to countries with less polluting technologies. This outcome is obtained because it is cheaper for economies with a cleaner technology to avoid emissions, so overall emissions in these countries should be lower. Furthermore, allowing for a nonlinear feedback effect of a higher average temperature in the climate subsystem, we could show that the economy may be characterized by multiple long-run balanced growth paths, implying that in this case initial conditions are crucial as to which equilibrium is obtained. Thus, there are tipping points where climate policies strongly matter to obtain desirable outcomes. This result is obtained on the basis of a carbon tax as a regulatory instrument which is preferred to a cap and trade system in our study.

The third part was concerned with economic growth and renewable and nonrenewable resources as well as with policies to prevent overextraction of those resources. With the currently ongoing process of global growth, there is a high demand for renewable and nonrenewable resources. This implies a strong externality effect across generations. The currently depleted resources are not available for future generations. For renewable and nonrenewable resources, we discussed the concept of sustainable growth and study how resource constraints can be overcome by substitution and technical progress. By building reasonable small-scale growth models for nonrenewable resources, we studied those issues and also estimated model variants and studied the time to exhaustion of specific resources.

Concerning renewable resources, we also explored small-scale dynamic decision models, which allowed us to analyze the fate of the resources when they are extracted. We were able to demonstrate that the usual results one obtains from the optimal exploitation of one resource do not carry over to ecologically interacting resources. Technically, we also showed how short and long horizon models hang together. We demonstrated how competition, in particular in a short time horizon context, leads to a faster depletion of resources. We also addressed the policy question of how regulatory instruments can be used to prevent the overextraction of natural resources. Although only tax rates are analyzed as regulatory instruments to prevent the depletion of the natural resources, we demonstrated that our approach lends itself to the study of other regulatory instruments.

It is worth reminding the reader that the issue of public regulation of the overextraction of natural resources was at the heart of the beginning of studies on natural resources.[1] As Hotelling (1931) pointed out, natural resources are not properly regulated under either free competition (which may lead to an overexploitation of resources) or monopoly (which may lead to high prices and monopoly profits). Some public regulation is needed. In part III of the book, we came to similar conclusions.

Essential for the externality effects on future generations—either the overuse of resources or pollution and climate change—is the size of the

---

[1] See, for example, the seminal paper by Hotelling (1931), using an intertemporal framework.

discount rate. Already Hotelling (1931) made a difference between the market operation for which it is reasonable to use a market rate of interest as discount rate and some resources of social value that may be valued higher (and thus discounted at a lower rate) than for the production of market goods. The size of the discount rate has also become crucial in the discussion on the Stern (2006, 2007) report, where it is argued that the almost zero discount rate, will overrate future damages arising from global warming and overstate current cost to make future damages less likely.[2] Yet, following Hotelling's distinction, it might make sense to suggest two different discount rates, one for resources of social value and one for market goods.

Overall, we presented a type of work that helps integrate the research on environmental and climate issues, as well as research on renewable and nonrenewable resources, into a consistent economic framework that takes the perspective of modern growth theory.

---

[2] For a detailed discussion on the issue of the discount rate see Nordhaus (2007a) and Weitzmann (2007c).

Appendix: Three Useful Theorems from Dynamic Optimization

In this book, we have presumed that economic agents behave intertemporally and perform dynamic optimization. In this appendix, we present some basics of the method of dynamic optimization using Pontryagin's maximum principle and the Hamiltonian.

Let an intertemporal optimization problem be given by

$$\max_{u(t)} W(x(0),0), W(\cdot) \equiv \int_0^\infty e^{-\rho t} F(x(t), u(t)) dt, \qquad (A.1)$$

subject to

$$\frac{dx(t)}{dt} \equiv \dot{x}(t) = f(x(t), u(t)), x(0) = x_0, \qquad (A.2)$$

with $x(t) \in \mathbb{R}^n$ the vector of state variables at time t and $u(t) \in \Omega \in \mathbb{R}^m$ the vector of control variables at time t and $F : \mathbb{R}^n \times \mathbb{R}^m \to \mathbb{R}$ and $f : \mathbb{R}^n \times \mathbb{R}^m \to \mathbb{R}^n$. ρ is the discount rate and $e^{-\rho t}$ is the discount factor.

$F(x(t), u(t))$, $f_i(x(t), u(t))$, and $\partial f_i(x(t), u(t))/\partial x_j(t)$, $\partial F(x(t), u(t))/\partial x_j(t)$ are continuous with respect to all $n + m$ variables for $i, j = 1, \ldots, n$ Further, $u(t)$ is said to be admissible if it is a piecewise continuous function on $[0, \infty)$ with $u(t) \in \Omega$.

Define the current-value Hamiltonian $\mathcal{H}(x(t), u(t), \lambda(t), \lambda_0)$ as follows:

$$\mathcal{H}(x(t), u(t), \lambda(t), \lambda_0) \equiv \lambda_0 F(x(t), u(t)) + \lambda(t) f(x(t), u(t)), \qquad (A.3)$$

with $\lambda_0 \in \mathbb{R}$ a constant scalar and $\lambda(t) \in \mathbb{R}^n$ the vector of co-state variables or shadow prices. $\lambda_j(t)$ gives the change in the optimal objective functional W^o resulting from an increment in the state variable $x_j(t)$. If $x_j(t)$ is a capital stock, $\lambda_j(t)$ gives the marginal value of capital at time t. Assume that there exists a solution for (A.1) subject to (A.2). Then, we have the following theorem.

Theorem A.1 *Let $u^o(t)$ be an admissible control and $x^o(t)$ is the trajectory belonging to $u^o(t)$. For $u^o(t)$ to be optimal, it is necessary that there*

194

exists a continuous vector function $\lambda(t) = (\lambda_1(t), \ldots, \lambda_n(t))$ *with piecewise continuous derivatives and a constant scalar* λ_0 *such that*

a. $\lambda(t)$ *and* $x^o(t)$ *are solutions of the canonical system*

$$\dot{x}^o(t) = \frac{\partial}{\partial \lambda} \mathcal{H}(x^o(t), u^o(t), \lambda(t), \lambda_0),$$

$$\dot{\lambda}(t) = \rho\lambda(t) - \frac{\partial}{\partial x} \mathcal{H}(x^o(t), u^o(t), \lambda(t), \lambda_0).$$

b. *For all* $t \in [0, \infty)$ *where* $u^o(t)$ *is continuous, the following inequality must hold:* $\mathcal{H}(x^o(t), u^o(t), \lambda(t), \lambda_0) \geq \mathcal{H}(x^o(t), u(t), \lambda(t), \lambda_0)$,

c. $(\lambda_0, \lambda(t)) \neq (0, 0)$ *and* $\lambda_0 = 1$ *or* $\lambda_0 = 0$.

Remarks:

1. If the maximum with respect to $u(t)$ is in the interior of Ω, $\partial\mathcal{H}(\cdot)/\partial u(t) = 0$ can be used as a necessary condition for a local maximum of $\mathcal{H}(\cdot)$.
2. It is implicitly assumed that the objective functional (A.1) takes on a finite value, that is, $\int_0^\infty e^{-\rho t} F(x^o(t), u^o(t)) < \infty$. If x^o and u^o grow without an upper bound $F(\cdot)$ must not grow faster than ρ.

Theorem A.1 provides only necessary conditions. The next theorem gives sufficient conditions.

Theorem A.2 *If the Hamiltonian with* $\lambda_0 = 1$ *is concave in* $(x(t), u(t))$ *jointly and if the transversality condition* $\lim_{t\to\infty} e^{-\rho t}\lambda(t)(x(t) - x^o(t)) \geq 0$ *holds, conditions a and b from theorem A.1 are also sufficient for an optimum. If the Hamiltonian is strictly concave in* $(x(t), u(t))$ *the solution is unique.*

Remarks:

1. If the state and co-state variables are positive the transversality condition can be written as stated in the foregoing chapters, that is, as $\lim_{t\to\infty} e^{-\rho t}\lambda(t)x^o(t) = 0$.[1]
2. Given some technical conditions, it can be shown that the transversality condition is also a necessary condition.

Theorem A.2 requires joint concavity of the current-value Hamiltonian in the control and state variables. A less restrictive theorem is the following.

Theorem A.3 *If the maximized Hamiltonian*

$$\mathcal{H}^o(x(t), \lambda(t), \lambda_0) = \max_{u(t) \in \Omega} \mathcal{H}(x(t), \lambda(t), \lambda_0)$$

---

[1] Note that in the book we did not indicate optimal values by ^{o}.

with $\lambda_0 = 1$ is concave in $x(t)$ and if the transversality condition $\lim_{t \to \infty} e^{-\rho t} \lambda(t)(x(t) - x^o(t)) \geq 0$ holds, conditions a and b from theorem A.1 are also sufficient for an optimum. If the maximized Hamiltonian $\mathcal{H}^o(x(t), \lambda(t), \lambda_0)$ is strictly concave in $x(t)$ for all t, $x^o(t)$ is unique (but not necessarily $u^o(t)$).

Because the joint concavity of $\mathcal{H}(x(t), u(t), \lambda(t), \lambda_0)$ with respect to $(x(t), u(t))$ implies concavity of $\mathcal{H}^o(x(t), \lambda(t), \lambda_0)$ with respect to $x(t)$, but the reverse does not necessarily hold, theorem A.3 may be applicable where theorem A.2 cannot be applied.

The three theorems demonstrate how optimal control theory can be applied to solve dynamic optimization problems. The main role is played by the Hamiltonian (A.3). It should be noted that in most economic applications, as in this book, interior solutions are optimal so that $\partial \mathcal{H}(\cdot) / \partial u(t) = 0$ can be presumed. For further reading and more details concerning optimal control theory, we refer to the books by Feichtinger and Hartl (1986) or Seierstad and Sydsaeter (1987).

Bibliography

Arrow, K. and M. Kurz (1970) *Public Investment, the Rate of Return, and Optimal Fiscal Policy*. Baltimore: John Hopkins Press.

Asada, T. and W. Semmler (1995) "Growth, Finance and Cycles: An Intertemporal Model." *Journal of Macroeconomics* 17(4): 623–649.

Aschauer, D.A. (1989) "Is Public Expenditure Productive?" *Journal of Monetary Economics* 23: 177–200.

Azar, C. and S.H. Schneider (2003) "Are the Economic Costs of (Non-)Stabilizing the Atmosphere Prohibitive? A Response to Gerlagh and Papyrakis." *Ecological Economics* 46: 329–332.

Barro, R.J. (1990) "Government Spending in a Simple Model of Endogenous Growth." *Journal of Political Economy* 98: S103–S125.

Becker, S., L. Grüne and W. Semmler (2007) "Comparing Accuracy of Second-order Approximation and Dynamic Programming." *Computational Economics* 30: 65–91.

Beckerman, W. (1974) *In Defence of Economic Growth*. London: Trinity Press.

Beltratti, A., G. Chichilnisky and G.M. Heal (1994) "The Environment and the Long-Run: A Comparison of Different Criteria." *Richerche Economiche* 48: 319–340.

Benhabib, J. and R.E.A. Farmer (1994) "Indeterminacy and Increasing Returns." *Journal of Economic Theory* 63: 19–41.

Benhabib, J. and K. Nishimura (1979) "The Hopf-Bifurcation and the Existence and Stability of Closed Orbits in Multisector Models of Optimal Growth." *Journal of Economic Theory* 21: 412–444.

Benhabib, J. and R. Perli (1994) "Uniqueness and Indeterminacy: On the Dynamics of Endogenous Growth," *Journal of Economic Theory* 63: 113–142.

Benhabib, J., Q. Meng and K. Nishimura (2000) "Indeterminacy under Constant Returns to Scale in Multisector Economies." *Econometrica* 68: 1541–1548.

Benhabib, J., R. Perli and D. Xie (1994) "Monopolistic Competition, Indeterminacy and Growth." *Ricerche Economiche* 48: 279–298.

Berck, P. (1979) "Open Access and Extinction." *Econometrica* 47(4): 877–883.

Boldrin, M. and L. Montrucchio (1986) "On the Indeterminacy of Capital Accumulation Paths." *Journal of Economic Theory* 40: 26–39.

Bovenberg, L.A. and R.A. de Mooij (1997) "Environmental Tax Reform and Endogenous Growth." *Journal of Public Economics* 63: 207–237.

Bovenberg, L.A. and S. Smulders (1995) "Environmental Quality and Pollution-Augmenting Technological Change in a Two-Sector Endogenous Growth Model." *Journal of Public Economics* 57: 369–391.

Brock, W.A. and J.A. Scheinkman (1976) "Global Asymptotic Stability of Optimal Control System with Applications to the Theory of Economic Growth." *Journal of Economic Theory* 12: 164–190.

Brock, W.A. and M.S. Taylor (2004) "Economic Growth and the Environment: A Review of Theory and Empirics." NBER Working Paper, 10854.

Broecker, W.S. (1997) "Thermohaline Circulation, the Achilles Heel of Our Climate System: Will Man Made CO_2 Upset the Current Balance?" *Science* 278: 1582–1588.

Brundtlandt Commission (1987) *Report of the World Commission on Environment and Development. Our Common Future.* United Nations.

Buonanno, P., C. Carraro and M. Galeotti (2003) "Endogenous Induced Technical Change and the Costs of Kyoto." *Resource and Energy Economics* 21: 11–34.

Byrne, M. (1997) "Is Growth a Dirty Word? Pollution, Abatement and Endogenous Growth." *Journal of Development Economics* 54: 261–284.

Carlson, D.A. and A. Haurie (1987) *Infinite Horizon Optimal Control, Theory and Applications.* Heidelberg: Springer. Lecture Notes in Economics and Mathematical Systems, vol. 290.

Cass, D. and K. Shell (1976) "The Structure and Stability of Competetive Systems." *Journal of Economic Theory* 12: 31–70.

Chamley, C. (1986) "Optimal Taxation of Capital Income in General Equilibrium with Infinite Lives." *Econometrica* 54(3): 607–622.

Chichilnisky, G. (1996) "What Is Sustainable Development?" Paper presented at Stanford Institute for Theortical Economics, 1993; published as "An Axiomatic Approach to Sustainable Development," *Social Choice and Wefare* 13(2): 219–248.

Citibase (1998). U.S. Basic Economics Database, Ball Sate University, Muncie, IN.

Clark, C.W. (1971) "Economically Optimal Policies for the Utilization of Biologically Renewable Resources." *Mathematical Bioscience* 17: 245–368.

——— (1985) *Bioeconomic Modelling and Fishery Management.* New York: Wiley Interscience.

——— (1990) *Mathematical Bioeconomics: The Optimal Management of Renewable Resources.* New York: J. Wiley. (first ed. 1976).

Clark, C.W., F.H. Clarke and G.R. Munro (1979) "The Optimal Exploitation of Renewable Resource Stocks: Problems of Irreversible Investment." *Econometrica* 47: 25–47.

Conrad, J.M. and R. Adu-Asamoah (1986) "Single and Multispecies Systems: The Case of Tuna in the Eastern Tropical Atlantic." *Journal of Environmental Economics and Management* 13: 50–86.

Daly, H.E. (1987) "The Economic Growth Debate: What Some Economists Have Learned But Many Have Not." *Journal of Environmental Economics and Management* 14(4): 323–336.

Dasgupta, P. and G. Heal (1974) "The Optimal Depletion of Exhaustible Resources." *Review of Economic Studies* (Symposium on the Economic of Exhaustible Resources): 3–28.

——— (1979) *Economic Theory and Exhaustible Resources.* Welwyn: Cambridge University Press.

Dechert, W.D. and K. Nishimura (1983) "A Complete Characterization of Optimal Growth Path in an Agregate Model with Non-Concave Production Function." *Journal of Economic Theory* 31: 332–354.

Deke O., K.G. Hooss, C. Kasten, G. Klepper and K. Springer (2001) "Economic Impact of Climate Change: Simulations with a Regionalized Climate-Economy Model." Kiel Working Paper, no. 1065.

Deutsch, C., M.G., Hall, D.F., Bradford and K., Keller, (2002) "Detecting a Potential Collapse of the North Atlantic Thermohaline Circulation: Implications for the Design of an Ocean Observation System." Mimeo Princeton University.

Dixit, A. (1989) "Entry and Exit Decisions under Uncertainty." *Journal of Political Economy* 97(3): 620–638.

Dockner, E.J. and G. Feichtinger (1991) "On the Optimality of Limit Cycles in Dynamic Economic Systems." *Journal of Economics* 53: 31–50.

Energy Information Administration. Official Energy Statistics from the U.S. Government (www.eia.doe.gov).

Falcone, M. (1987) "A Numerical Approach to the Infinite Horizon Problem of Deterministic Control Theory." *Applied Mathematics and Optimization* 15: 1–13.

Falk, I. (1988) "A Dynamic Model of Interrelated Renewable Resources." *Resources and Energy* 10: 55–77.

Fankhauser, S. (1995) *Valuing Climate Change. The Economics of the Greenhouse Effect*. London: Earthscan.

Feichtinger, G. and R.F. Hartl (1986) *Optimale Kontrolle Ökonomischer Prozesse: Anwendungen des Maximumprinzips in den Wirtschaftswissenschaften*. Berlin: de Gruyter.

Feichtinger, G., V. Kaitala and A. Novak (1991) "Stable Resource-Employment Limit Cycles in an Optimally Regulated Fishery." Mimeo, University of Technology, Vienna.

Forster, B.A. (1973) "Optimal Capital Accumulation in a Polluted Environment." *Southern Economic Journal* 39: 544–547.

Futagami, K., Y. Morita and A. Shibata (1993) "Dynamic Analysis of an Endogenous Growth Model with Public Capital." *Scandinavian Journal of Economics* 95: 607–625.

Gassmann, F. (1992) "Die wichtigsten Erkenntnisse zum Treibhaus-Problem." In Schweizerische Fachvereeinigung für Energiewirtschaft, ed., *Wege in eine CO_2–arme Zukunft*. Zürich: Verlag der Fachvereine, pp. 11–25.

Gerlagh, R. (2004) "Climate Change and Induced Technological Change." Paper presented at the 13th Annual Conference of the European Association of Environmental and Resource Economics (EAERE). Available at eaere2004.bkae.hu/download/paper/gerlaghpaper.pdf.

Gerlagh, R. and E. Papyrakis (2003) "Are the Economic Costs of (Non-) Stabilizing the Atmosphere Prohibitive? A Comment." *Ecological Economics* 46: 325–327.

Gordon, F.S. (1954) "The Economic Theory of a Common Property Resource: The Fishery." *Journal of Political Economy* 62: 124–138.

Gradus, R. and S. Smulders (1993) "The Trade-Off between Environmental Care and Long-Term Growth—Pollution in Three Prototype Growth Models." *Journal of Economics* 58: 25–51.

Greiner A. (2004a) "Anthropogenic Climate Change in a Descriptive Growth Model." *Environment and Development Economics* 9: 645–662.

Greiner A. (2004b) "Global Warming in a Basic Endogenous Growth Model."
    Environmental Economics and Policy Studies 6: 49–73.
———— (2005a) "Fiscal Policy in an Endogenous Growth Model with Public
    Capital and Pollution." *Japanese Economic Review* 56: 67–84.
———— (2005b) "Anthropogenic Climate Change and Abatement in a Multi-
    Region World with Endogenous Growth." *Ecological Economics* 55: 224–234.
———— (2007) "The Dynamic Behaviour of an Endogenous Growth Model
    with Public Capital and Pollution." *Studies in Nonlinear Dynamics and
    Econometrics* 11.
Greiner, A. and W. Semmler (2005) "Economic Growth and Global Warming:
    A Model of Multiple Equilibria and Thresholds." *Journal of Economic
    Behaviour and Organization* 57: 430–447.
Greiner, A., W. Semmler and G. Gong (2005) *The Forces of Economic Growth—
    A Time Series Approach*. Princeton: Princeton University Press.
Grüne, L. and W. Semmler (2004) "Using Dynamic Programming with Adap-
    tive Grid Scheme for Optimal Control Problems in Economics." *Journal of
    Economic Dynamics and Control* 28: 2427–2456.
Grüne, L., M. Kato and W. Semmler (2005) "Solving Ecological Management
    Problems Using Dynamic Programming." *Journal of Economic Behavior and
    Organization* 57(4): 448–474.
Gruver, G. (1976) "Optimal Investment and Pollution Control in a Neoclassi-
    cal Growth Context." *Journal of Environmental Economics and Management* 5:
    165–177.
Hackl, F. and G.J. Pruckner (2003) "How Global Is the Solution to Global
    Warming?" *Economic Modelling* 20: 93–117.
Hannesson, R. (1983) "Optimal Harvesting of Ecologically Interrelated Fish
    Species." *Journal of Environmental Economics and Management* 10: 329–345.
Harvey, D.L.D. (2000) *Global Warming—The Hard Science*. Harlow: Prentice Hall.
Haurie, A. (2003) "Integrated Assessment Modeling for Global Climate Change:
    An Infinite Horizon Optimization Viewpoint." *Environmental Modeling and
    Assessment* 2: 117–132.
Heal, G. (1999) *Valuing the Future: Economic Theory and Sustainability*. New York:
    Columbia University Press.
Heal. G.M. (1993) "The Relationship between Price and Extraction Cost for a
    Resource with a Backstop Technology." In Geoffrey Heal, ed., *The Economics of
    Exhaustible Resources*. International Library of Critical Writings in Economics.
    Cheltenham: Edward Elgar.
———— (1995) "Interpreting Sustainability." *Nota Di Lavoro* 1: 1–14.
Henderson-Sellers, A. and K. McGuffie (1987) *A Climate Modelling Primer*.
    Chichester: John Wiley.
Hettich, F. (1998) "Growth Effects of a Revenue-Neutral Environmental Tax
    Reform." *Journal of Economics* 67: 287–316.
———— (2000) *Economic Growth and Environmental Policy*. Cheltenham: Edward
    Elgar.
Hirsch, M.W. and S. Smale (1974) *Differential Equation, Dynamical Systems and
    Linear Algebra*. New York: Academic Press.
Hotelling, H. (1931) "The Economics of Exhaustible Resources." *Journal of
    Political Economy* 39: 137–175.
Institut der deutschen Wirtschaft (2003) *Deutschland in Zahlen*. Köln.

IPCC (1990) *Climate Change. The IPCC Scientific Assessment.* Cambridge: Cambridge University Press.

—— (1996) "Climate Change 1995: Economic and Social Dimensions of Climate Change." Contribution of Working Group III to the Second Assessment Report of the IPCC. In J.P. Bruce, H. Lee and E.F. Haites, eds. *Economic and Social Dimensions of Climate Change.* Cambridge: Cambridge University Press, pp. 40–78.

—— (2001) "Climate Change 2001: The Scientific Basis." IPCC Third Assessment Report of Working Group I. Available at www.ipcc.ch.

Jones, L., R.E. Manuelli and P.E. Rossi (1993) "Optimal Taxation in Models of Economic Growth." *Journal of Political Economy* 101(3): 485–517.

Judd, K.L. (1987) "The Welfare Cost of Factor Taxation in a Perfect Foresight Model." *Journal of Political Economy* 49(4): 675–707.

Kalecki, M. (1938) *Essays on Economic Dynamics.* Cambridge: Cambridge University Press (1966 edition).

Keller, K., K. Tan, F.M. Morel and D.F. Bradford (2000) "Preserving the Ocean Circulation: Implications for the Climate Policy." *Climate Change* 47: 17–43.

Kemfert, C. (2001) *Economy-Energy-Climate Interaction. The Model WIAGEM.* Fondazione Eni Enrico Mattei, Nota di Lavoro (FEEM Working Paper) 71.2001.

King, R.G. and S. Rebelo (1990) "Public Policy and Economic Growth: Developing Neoclassical Implications." *Journal of Political Economy* 98(5): 126–150.

Koopmans, T.C. (1985) "The Transition from Exhaustible to Renewable or Inexhaustible Resources." In T.C. Koopmans, *Scientific Papers*, vol. 2. Cambridge, MA: MIT Press.

Koskela, E., M. Ollikainen and M. Puhakka (2000) "Saddles, Indeterminacy, and Bifurcations in an Overlapping Generations Economy with a Renewable Resource." CESifo, Working Paper Series no. 298.

Krautkraemer, J.A. (1985) "Optimal Growth, Resource Amenities and the Preservation of Natural Environments." *Review of Economic Studies* 52: 153–171.

Ligthart, J.E. and F. van der Ploeg (1994) "Pollution, the Cost of Public Funds, and Endogenous Growth." *Economics Letters* 46: 339–348.

Lucas, R.E. (1988) "On the Mechanics of Economic Growth." *Journal of Monetary Economics* 22: 3–42.

—— (1990) "Supply-Side Economics: An Analytical Review." *Oxford Economic Papers* 42: 293–316.

Luptacik, M. and U. Schubert (1982) *Optimal Economic Growth and the Environment: Economic Theory of Natural Resources.* Vienna: Physica.

Majumdar, M. and T. Mitra (1982) "Inter-temporal Allocation with Nonconvex Technology, the Aggregative Framework." *Journal of Economic Theory* 27: 101–136.

Mäler, K.G. (1974) *Environmental Economics: A Theoretical Inquiry.* Baltimore: Johns Hopkins University Press.

Meadows, D.H., D.L. Meadows, J. Randers and W.W. Behrens (1972) *The Limits to Growth.* New York: Universe Books.

Montrucchio, L. (1992) "Dynamical Systems That Solve Continuous Time Concave Optimization Problems: Anything Goes." In J. Benhabib, ed., *Cycles and Chaos in Economic Equilibrium*, Princeton: Princeton University Press.

Nielsen, S.B., L.H. Pedersen and P.B. Sorensen (1995) "Environmental Policy, Pollution, Unemployment, and Endogenous Growth." *International Tax and Public Finance* 2: 185–205.

Nordhaus W.D. (1994) *Managing the Global Commons: The Economics of Climate Change.* Cambridge, MA: MIT Press.

——— (2007a) "The Stern Review on the Economics of Climate Change." *Journal of Economics Literature* 45: 686–702.

——— (2007b) *The challenge of Global Warning: Economic Models and Environmental Policy.* New Haven: Yale University Press.

Nordhaus, W.D. and J. Boyer (2000) *Warming the World. Economic Models of Global Warming.* Cambridge, MA: MIT Press.

Nyarko, Y. and L.J. Olson (1996) "Optimal Growth with Unobservable Resources and Learning." *Journal of Economic Behavior and Organization* 29: 465–491.

OECD (1995) "Global Warming—Economic Dimensions and Policy Responses." Paris.

Pearce, D., E.B. Barbier and A. Markandya (1990) *Sustainable Development: Economics and Environment in the Third World.* Worcester: Edward Elgar.

Peck, S. and T.J. Teisberg (1992) "CETA: A Model for Carbon Emissions Trajectory Assessment." *Energy Journal* 13: 55–77.

Pezzey, E.S. (1961) "Economic Analysis of Sustainable Growth and Sustainable Development." World Bank Policy Planning and Research Staff, Environment Department Working Paper no. 15.

Pfähler, W., U. Hofmann and W. Bönte (1996) "Does Extra Public Capital Matter? An Appraisal of Empirical Literature." *Finanzarchiv N.F.* 53: 68–112.

Phelps, E.S. (1961) "The Golden Rule of Accumulation: A Fable for Growthmen." *American Economic Review* 638–643.

Pindyck, R.S. (1991) "Irreversibility, Uncertainty, and Investment." *Journal of Economics Literature* 29(3): 1110–1148.

Popp, D. (2003) "ENTICE: Endogenous Technological Change in the DICE Model of Global Warming." NBER Working Paper no. 9762.

Ragozin, D.L. and G. Brown (1985) "Harvest Policies and Nonmarket Valuation in a Predator-Prey System." *Journal of Environmental Economics and Management* 12: 155–168.

Ramsey, F. (1928) "A Mathematical Theory of Saving." *Economic Journal* 38: 543–559.

Rawls, J. (1972) *A Theory of Justice.* Oxford: Clarendon.

Rebelo, S. (1991) "Long-Run Analysis and Long-Run Growth." *Journal of Political Economy* 99(31): 500–521.

Rockafellar, R.T. (1976) "Saddle Points of Hamiltonian Systems in Convex Lagrange Problems Having a Nonzero Discount Rate." *Journal of Economic Theory* 12: 71–113.

Roedel, W. (2001) *Physik unserer Umwelt: Die Atmosphäre.* Berlin: Springer-Verlag.

Romer, P.M. (1986) "Increasing Returns and Long-Run Growth." *Journal of Political Economy* 94: 1002–1037.

——— (1990) "Endogenous Technological Change." *Journal of Political Economy* 98: S71–S02.

Romp, W. and J. de Haan (2005) "Public Capital and Economic Growth: A Critical Survey." *EIB Papers* 10 (1).

Rosen, J.B. (1965) "Existence and Uniqueness of Equilibrium Points for Concave N-Person Games." *Econometrica* 33: 520–534.

Sachverständigenrat zur Begutachtung der gesamtwirtschaftlichen Lage (2001), *Chancen auf einen höheren Wachstumspfad.* Stuttgart: Metzler-Poeschel.

Schmitz, G. (1991) "Klimatheorie und -modellierung." In P. Hupfer, ed., *Das Klimasystem der Erde: Diagnose und Modellierung, Schwankungen und Wirkungen.* Berlin: Akademie Verlag, pp. 181–217.

Scholl, A. and W. Semmler (2002) "Susainable Economic Growth and Exhaustible Resources: A Model and Estimation for the US." *Discrete Dynamics in Nature and Society* 7(2): 79–92.

Seierstad, A. and K. Sydsaeter, (1987) *Optimal Control with Economic Applications.* Amsterdam: North-Holland.

Semmler, W. (1994) "On the Optimal Regulation of an Extractive Industry." *Journal of Business and Economics* 46: 409–420.

—— (1994b) "On the Optimal Exploitation of Interacting Resources." *Journal of Economics* 59: 23–49.

—— (2000) "Debt Dynamics and Sustainable Debt." *Journal of Economic Dynamics and Control* 24: 1121–1144.

Semmler, W., A. Greiner, B. Dialo, A. Rezai and A. Rajaram (2007) "Fiscal Policy, Public Expenditure Composition, and Economic Growth." World Bank Policy Research working paper no. 4405, Washington, DC.

Sieveking, M. (1990) "Stability of Limit Cycles." Mimeo, Department of Mathematics, University of Frankfurt.

Sieveking, M. and W. Semmler (1990) "Optimization without Planning: Growth and Resource Exploitation with a Discount Rate Tending to Infinity." Working Paper, New School for Social Research.

—— (1997) "The Present Value of Resources with Large Discount Rates." *Applied Mathematics and Optimization* 38: 283–309.

Smith, V. (1968) "Economics of Production from Natural Resources." *American Economic Review* 58: 4909–4931.

Smulders, S. (1995) "Entropy, Environment, and Endogenous Growth." *International Tax and Public Finance* 2: 319–340.

Smulders, S., Gradus S. (1996) "Pollution Abatement and Long-Term Growth." *European Journal of Political Economy* 12: 505–532.

Solow, R.M. (1973) "Is the End of the World at Hand?" In Andrew Weintraub, Eli Schwartz, and J. Richard Aronson, eds., *The Economic Growth Controversy.* New York: International Arts and Sciences Press.

—— (1974) "Intergenerational Equity and Exhaustible Resources." *Review of Economic Studies* (Symposium on the Economics of Exhaustible Resources): 29–45.

Sorger, G. (1989) "On the Optimality and Stability of Competitive Paths in Time Continuous Growth Models." *Journal of Economic Theory* 48: 526–547.

Statistisches Bundesamt (2000) *Statistisches Jahrbuch 2000 für die Bundesrepublik Deutschland.* Metzler-Poeschel, Stuttgart.

Stern, N. (2006, 2007) "What Is the Economic Impact of Climate Change? Stern Review on the Economics of Climate Change." Discussion Paper, www.hm-treasury.gov.uk; printed version (2007) Cambridge: Cambridge University Press.

Stiglitz, J.E. (1974) "Growth with Exhaustible Natural Resources: Efficient and Optimal Growth Paths." *Review of Economic Studies* (Symposium on the Economics of Exhaustible Resources): 123–137.

Stockey, N.L. (1998) "Are There Limits to Growth?" *International Economic Review* 39: 1–31.

Sturm, J.E., G.H. Kuper and J. de Haan (1998) "Modelling Government Investment and Economic Growth on a Macro Level." In S. Brakman, H. van Ees and S.K. Kuipers, eds., *Market Behaviour and Macroeconomic Modelling*, London: Macmillan/St. Martin's Press, pp. 359–406.

Tol, R.S.J. (1999) "Spatial and Temporal Efficiency in Climate Policy: An Application of FUND." *Environmental and Resource Economics* 14: 33–49.

——— (2001) "Equitable Cost-Benefit Analysis of Climate Change." *Ecological Economics* 36: 71–85.

——— (2003) "Is the Uncertainty about Climate Change Too Large for Expected Cost-Benefit Analysis." *Climatic Change* 56: 265–289.

Toman, M.A., J. Pezzey and J.A. Krautkraemer (1993) "Neoclassical Economic Growth Theory and Sustainability." In Daniel W. Bromley, ed., *Handbook of Enviroenmental Economics*. Cambridge, MA: Blackwell Press.

U.S. Department of Commerce, Economics and Statistics, Bureau of the Cencus (1965–1997) *Statistical Abstract of the United States*.

U.S. Geological Survey (1995) Maps, Imaging and publications. http://www.usgs.gov/

Uzawa H. (2003) *Economic Theory and Global Warming*. New York: Cambridge University Press.

Weitzman, M. (2007a) "The Role of Uncertainty in the Economics of Catastrophic Climate Change." Manuscript, MIT.

——— (2007b) "Subjective Expectations and Asset Return Puzzles." Manuscript, MIT.

——— (2007c) "A Review of the Stern Review on the Economics of Climate Change." *Journal of Economics Literature* 45:703–25.

Weizsäcker, C.C. (1967) "Symposium on Optimal Infinite Programmes: Lemmas for a Theory of Approximate Optimal Growth." *Review of Economic Studies* 34: 143–151.

Index